Opportunities and Risks of 5G Military Use in Europe

MARY LEE, JAMES DIMAROGONAS, EDWARD GEIST, SHANE MANUEL,
RYAN A. SCHWANKHART, BRYCE DOWNING

Prepared for the Office of the Under Secretary of Defense for Research
and Engineering
Approved for public release; distribution unlimited

NATIONAL DEFENSE RESEARCH INSTITUTE

For more information on this publication, visit **www.rand.org/t/RRA1351-2**.

About RAND

The RAND Corporation is a research organization that develops solutions to public policy challenges to help make communities throughout the world safer and more secure, healthier and more prosperous. RAND is nonprofit, nonpartisan, and committed to the public interest. To learn more about RAND, visit www.rand.org.

Research Integrity

Our mission to help improve policy and decisionmaking through research and analysis is enabled through our core values of quality and objectivity and our unwavering commitment to the highest level of integrity and ethical behavior. To help ensure our research and analysis are rigorous, objective, and nonpartisan, we subject our research publications to a robust and exacting quality-assurance process; avoid both the appearance and reality of financial and other conflicts of interest through staff training, project screening, and a policy of mandatory disclosure; and pursue transparency in our research engagements through our commitment to the open publication of our research findings and recommendations, disclosure of the source of funding of published research, and policies to ensure intellectual independence. For more information, visit www.rand.org/about/principles.

RAND's publications do not necessarily reflect the opinions of its research clients and sponsors.

About This Report

This report focuses on U.S. military uses of the fifth-generation (5G) technology standard for cellular communications in a notional 2030 time frame, concentrating on a future *smart logistics* mission in the Baltics and surrounding countries. The report describes research conducted between April 2021 and February 2022 for the RAND Corporation's National Defense Research Institute on *Operational Risks and Opportunities of European 5G Use*, sponsored by the Operate Through Program Lead of the 5G-to-NextG Initiative at the Office of the Under Secretary of Defense for Research and Engineering. It is intended for policy-minded readers and decisionmakers and should be of interest to Joint decisionmakers with responsibilities for communications, emerging technologies, and logistics.

The research reported here was completed in October 2022 and underwent security review with the sponsor and the Defense Office of Prepublication and Security Review before public release.

RAND National Security Research Division

This research was sponsored by the Office of the Under Secretary of Defense for Research and Engineering and conducted within the Acquisition and Technology Policy Center of the RAND National Security Research Division (NSRD), which operates the National Defense Research Institute (NDRI), a federally funded research and development center sponsored by the Office of the Secretary of Defense, the Joint Staff, the Unified Combatant Commands, the Navy, the Marine Corps, the defense agencies, and the defense intelligence enterprise.

For more information on the RAND Acquisition and Technology Policy Center, see www.rand.org/nsrd/atp or contact the director (contact information is provided on the webpage).

Acknowledgments

The authors thank project sponsor Dan Massey for his guidance and feedback during this research. We also thank project monitors Katherine Borowec, Tammi Fisher, and Jennifer Peyrot. We are grateful to Joel Predd, Caitlin Lee, Yun Kang, and Chris Mouton for their program management. We thank RAND colleagues Brien Alkire, Tim Bonds, Jon Fujiwara, Dan Gonzales, Jeff Hagen, Tom Hamilton, Gavin Hartnett, Michael Kennedy, Sherrill Lingel, Mike Linick, Nick O'Donoughue, Chad Ohlandt, Osonde Osoba, and Lloyd Thrall for their expertise and feedback. We are grateful to Tim Bonds, Jim Powers, and Mitchell Stevenson for their careful and thoughtful reviews of this document.

Summary

Issue

The fifth-generation (5G) technology standard for broadband cellular communications is expanding in Europe and will offer many more capabilities than the existing fourth generation long-term evolution standard. With this increase in capabilities comes opportunities for the U.S. Department of Defense (DoD) to integrate advanced technologies and improved communications into its operations. However, these opportunities come with inherent risks. This research aimed to describe the 5G rollout in Europe, characterize what Russian experts have determined regarding the military utility of 5G, and identify DoD opportunities and risks of using the 5G ecosystem in a future Baltics scenario.

Approach

The research involved literature and document reviews of the 5G rollout and threats to 5G in countries of interest. We also conducted a literature review of primary sources, including Russian-language sources, to assess Russian thoughts on 5G. We developed a *smart logistics* vignette to evaluate the benefits of 5G using a consensus of 11 subject-matter experts (SMEs) on three aspects of 5G across a variety of tasks during the vignette: operational impact of 5G, resilience with 5G, and uniqueness of 5G. Using these reviews, assessments, and SME consensus, we focused on characterizing the 5G rollout in Europe and identifying the risks and benefits of the military's use of 5G in the European theater.

Key Findings

Communications are necessary for almost every aspect of warfighting, so it is no surprise that as 5G is rolled out, we can expect to see adversarial threats to 5G networks. Past and theorized threats imply that 5G could be at risk to man-in-the-middle attacks, exploitation via the lack of authentication requirements, or inherited vulnerabilities from existing cellular networks. Russian jammers have been or can be deployed; they are particularly deceptive because their operations can be disguised as everyday fixtures, such as ordinary transport vehicles. Jammers can be used to disrupt communications or deliver disinformation to blue forces.

In our analysis of Russian thoughts on 5G, we found the following:

- The trajectory of 5G in Russia is uncertain because of the hesitation of the Russian military to give up its rights to the contested 3.4 to 3.8 GHz band.
- Russia's 5G regulatory problems pose a challenge for such countries as Lithuania that are trying to cultivate 5G networks. North Atlantic Treaty Organization (NATO)

governments feel the need to abide by international agreements governing spectrum usage in peacetime, and NATO militaries cannot take advantage of 5G infrastructure that remains unbuilt because of regulatory conflicts with the Russians.

- If 5G technologies play some role in a conflict with Russia, Russia will attempt to counter 5G technologies in accordance with Russia's well-cultivated, holistic approach to *radioelectronic struggle* (*radioelektronnaia bor'ba*) or *electronic warfare* (EW). If these EW measures prove effective, the electromagnetic spectrum will potentially be a nonpermissive environment for 5G technologies during a conflict with Russia.

Despite these threats, 5G and next-generation cellular communications can help augment U.S. military operations. We found that four main functions emerged as 5G benefits:

- the ability to track thousands of items using smart tags—these inexpensive tags can help track items and equipment, and those data can be used in artificial intelligence or maintenance algorithms to enhance operations
- very high communications speeds over short distances with high-band 5G—this can help with downloading large sensor data sets (for example, to transfer data between operators and autonomous vehicles)
- remote communications—5G can be used for communications for command and control and allow for increased coordination among multinational forces and infrastructure
- unmanned communications—5G can allow for enhanced communications between autonomous vehicles.

The team found that, generally, SMEs did not have a favorable opinion of the uniqueness of 5G because many opportunities highlighted here can be achieved without 5G. However, the SMEs agreed that the four benefits listed above had high operational utility and resilience against red attack.

Recommendations

We recommend that the Office of the Under Secretary of Defense for Research and Engineering consider the following courses of action to realize the opportunities and mitigate the risks of military use of 5G in the European theater:

- Focus future 5G research on smart tags, high-band 5G, remote communications, and unmanned communications.
- Expand 5G research into different operational scenarios and vignettes, such as urban operations, humanitarian assistance and disaster recovery, and special forces operations.
- Develop mitigation approaches for countering Russian exploitation and denial of 5G.
- Track frequency developments across U.S. allies and their borders with Russia.
- Assess potential interoperability issues across providers and countries.
- Work with European allies to create a common roadmap of military uses of 5G.

Contents

Figures and Tables

Figures

Tables

Chapter 1. Introduction

The potential for military conflict breaking out between the United States, North Atlantic Treaty Organization (NATO) allies, and Russia is ever increasing. Operations by U.S. and NATO military forces in a future European wartime scenario will require exquisite communication and coordination across echelons. The fifth-generation (5G) cellular communications standard might help enable and promote the use of advanced technologies, such as unmanned and autonomous systems, artificial intelligence, and internet of things (IoT) devices, which will in turn alter future force projection and deployment of forces. 5G might thus offer valuable operational benefits, particularly when considering wartime logistics, i.e., planning, managing, and moving enormous numbers of personnel, supplies, and pieces of equipment. However, these benefits come with accompanying risks. Russia has demonstrated advanced capabilities in its use of the electromagnetic spectrum (EMS) to conduct electronic warfare (EW) and electronic attack. Such capabilities might reduce the benefits we might see with 5G. The research described in this report documents the potential opportunities and risks of using 5G in such a scenario.

What Is 5G?

5G can be a nebulous term, so we begin by describing it. Cellular networks have a list of requirements they must meet in terms of both the actual attributes of the network and the standard that the cellular networks must use to qualify as 5G. These requirements are laid out by the 3rd Generation Partnership Project (3GPP), a global partnership among seven telecommunications standards development organizations, in their Technical Specification (TS) 22.261 standard.[1] The 5G requirements include, among other things, three broad categories of requirements relevant to military operations: interoperability, network slicing, and performance.

In terms of interoperability, all 5G networks are required to support subscription roaming with all other 5G networks. It should be noted that this only requires the ability for such roaming to occur, not provision to any given user. Legacy support is also required to be considered 5G; specifically, networks must support all evolved packet system capabilities with some exceptions in some specific legacy services as described in the TS 22.261 standard.[2]

5G networks are also required to support network slicing, although the requirements for it are fairly lax. *Network slicing* is the ability to divide a single physical network into multiple logical

---

[1] 3GPP, *Service Requirements for the 5G System (3GPP TS 22.261 version 17.10.0 Release 17)*, Sophia Antipolis, France: ETSI, May 2022.

[2] 3GPP, 2022.

virtual networks, completely transparent to the users on the network.[3] Users on different slices are completely oblivious to other slices; it is as if they exist on separate physical networks. This allows for the dynamic provisioning of different networks with different characteristics while maintaining separation and security between these networks. There is no minimum required number of slices, but network operators must be able to create, delete, or modify slices. Specifically, network operators should be able to specify the prioritization, capacity, and functionality of slices in an ad hoc capacity. Related to the interoperability requirements, network slices are required to have the ability to support both home and roaming users on the same slice.

The key performance indicators set by the 3GPP include a peak data rate greater than 10 Gbps, 1 million connections per square km, and less than 1 ms latency for ultrareliable low-latency communications.[4] There are also performance requirements for supporting low-density areas. Specifically, in low-density areas with under two users per square km, 5G needs to support a download speed of 1 Mbps and upload speed of 100 kbps at the edge of this coverage, with a minimum of 10 Mbps for cellular phones.[5]

Frequency bands vary by country, but generally 5G has three bands: low band (700 MHz), mid band (3.6 GHz [3.4 to 3.8 GHz]), and high band (26 GHz [24.25 to 27.5 GHz]).[6] Industry is making large investments in the 26 GHz band, bringing down prices of the underlying electronics and antennas. This offers the U.S. Department of Defense (DoD) cost saving opportunities along with additional capabilities and functionality.

Capabilities Offered by 5G

5G promises a variety of capabilities that augment or exceed those seen in existing cellular and communications networks. Note that not all capabilities were stressed in our logistics vignette, described later, and hence were not examined in this work; however, they might be useful in other operational contexts.

---

[3] Qiang Chen, Xiaolei Wang, and Yingying Lv, "An Overview of 5G Network Slicing Architecture," *AIP Conference Proceedings*, Vol. 1967, No. 1, 2018.

[4] Eiman Mohyeldin, "Minimum Technical Performance Requirements for IMT-2020 Radio Interface(s)," presented at the ITU-R Workshop on IMT-2020 Terrestrial Radio Interfaces Evaluation, Geneva, Switzerland, December 2019.

[5] 3GPP, 2022.

[6] European 5G Observatory, "To Fulfil Its Potential 5G Needs Access to Much Higher Frequencies: 3.5 GHz and Above. This Was Not the Case with Earlier Mobile Generations," webpage, last updated 2021d.

5G High-Band Communications

5G high-band communications will provide high bandwidth and low-latency communications at the 26.5 to 29.5 GHz and 37 to 40 GHz frequencies.[7] At such high frequencies, 5G requires small antennas with narrow pencil beam communications. This will make it difficult to jam unless the adversary is right inside the beam and will also make it difficult to detect and geolocate the user because of the pencil beam communications' small sidelobes. Caveats to high-band 5G include the need for good line of sight and the fact that high-band 5G is limited in range. One use for 5G high-band communications might be to take advantage of its low latency for coordinated unmanned aircraft system swarm attacks and defenses.

5G for High-Capacity and/or Low-Latency Communications

High-capacity and low-latency communications are not restricted only to high-band 5G; other 5G frequency bands can also offer this feature.[8] The high bandwidth of 5G can allow faster downloads of massive data sets (for example, downloaded sensor data from F-35s). Ultrareliable low latency offered by 5G can enable applications to include such functions as remote surgery, factory automation, high speed vehicle coordination.

5G for Internet of Things

5G can support 1 million users per square kilometer,[9] which allows small and inexpensive smart tags to be used to keep track of objects and collect data from them.[10] With expected 5G speeds and bandwidth, such tags will be able to connect with each other and share data at volumes and speeds not yet seen.[11]

---

[7] Jaehun Lee, Ji-Seon Paek, and Songcheol Hong, "Frequency Reconfigurable Dual-Band CMOS Power Amplifier for Millimeter-Wave 5G Communications," *2021 IEEE MTT-S International Microwave Symposium (IMS)*, 2021.

[8] Julie Song, "Why Low Latency (Not Speed) Makes 5G a World-Changing Technology," *Forbes*, February 6, 2020.

[9] International Telecommunication Union Radiocommunication Sector, *Minimum Requirements Related to Technical Performance for IMT-2020 Radio Interface(s)*, Geneva, Switzerland, M.2410-0, November 2017, p. 6.

[10] Sotirios K. Goudos, Panagiotis I. Dallas, Stella Chatziefthymiou, and Sofoklis Kyriazakos, "A Survey of IoT Key Enabling and Future Technologies: 5G, Mobile IoT, Sematic Web and Applications," *Wireless Personal Communications*, Vol. 97, 2017.

[11] Goudos et al., 2017.

5G for Passive Sensing

5G signals can also be used for passive sensing by placing multiple 5G emitters at different frequencies.[12] 5G receivers can then be used to detect echoes of targets (whether ground, air, or ship). This could be particularly useful for small drone detection because drone propellers create distinct doppler signatures. 5G at higher frequencies offer even better location precision.

5G for Alternative Positioning, Navigation, and Timing

5G signals can be used to calculate position in Global Positioning System (GPS)-denied environments.[13] These provide a few advantages over 4G. One is that higher carrier frequencies have higher position accuracy, so there are fewer issues with multipath (which occurs when a signal propagates and arrives at a receiver via more than one path, which can result in interferences or distortion of the signal). The larger bandwidth also results in reduced multipath effects. Another advantage over 4G is that 5G allows more emitters from which to geolocate, which allows for better geometric diversity. 5G also offers higher receive power than GPS signals, which makes 5G signals difficult to jam.

5G Security and Network Slicing

In 5G, separate virtual networks (*slices*) can be created that ride on top of the same infrastructure.[14] Each slice can be provisioned with specific performance parameters, and slices would be prevented from monitoring or interfering with each other.[15] Thus, a malicious actor flow can be isolated in a separate slice to be monitored and analyzed without interfering with the rest of traffic.[16]

---

[12] Piotr Samczyński, Karol Abratkiewicz, Marek Płotka, Tomasz P. Zieliński, Jacek Wszołek, Sławomir Hausman, Piotr Korbel, and Adam Księżyk, "5G Network-Based Passive Radar," *IEEE Transactions on Geoscience and Remote Sensing,* Vol. 60, No. 5108209, 2022.

[13] Ali A. Abdallah, Kimia Shamaei, and Zaher M. Kassas, "Assessing Real 5G Signals for Opportunistic Navigation," *Proceedings of the 33rd International Technical Meeting of the Satellite Division of the Institute of Navigation (ION GNSS+ 2020),* September 2020.

[14] Daifallah Alotaibi, "Survey on Network Slice Isolation in 5G Networks: Fundamental Challenges," *Procedia Computer Science,* Vol. 182, 2021.

[15] Latif U. Khan, Ibrar Yaqoob, Nguyen H. Tran, Zhu Han, and Choong Seon Hong, "Network Slicing: Recent Advances, Taxonomy, Requirements, and Open Research Challenges," *IEEE Access,* Vol. 8, 2020.

[16] Note that although this can certainly be the case, network slicing also brings new security challenges that would need to be addressed. See Alotaibi, 2021, and Khan et al., 2020, for an in-depth discussion of those challenges and how they might be addressed.

Objective, Scope, and Limitations

The overarching objective of the research documented in this report is to outline the future operational risks and benefits of using 5G and next-generation cellular communications in a specific, idealized Baltics scenario. Three subtasks formed the basis of this work:

1. Baseline the European 5G network in the future. Nominally, we have considered the projected 5G network in the year 2030.
2. Review Russian use of 5G and implications for U.S., partner, and allied militaries.
3. Identify joint operational opportunities and risks to 5G use in a Baltics scenario.

This research is not meant to be a predictive analysis; the actual rollout and capabilities of 5G might differ from what is expected, which might alter the recommendations in this report.

Approach

The research team conducted reviews of open-source literature, including Russian-language sources where applicable and available, to baseline the 5G network, assess implications of Russian use of 5G, and identify operational risks. The team also held a SME workshop with 11 participants to gather qualitative evaluations of 5G operational impacts. Each participant had expertise in one or more of the following areas:

- military operations (sustainment and logistics)
- technical expertise (communications, including 5G)
- autonomous vehicles, including unmanned aircraft systems
- artificial intelligence
- EW and cyber.

Participants were asked to qualitatively score components of the vignette on the following attributes: operational impact of 5G, resilience with 5G, and 5G uniqueness. More details about the SME workshop are provided in Chapter 5. The vignette is summarized next, and a full description is in Appendix A.

Summary of Smart Logistics Vignette

The team developed a smart logistics vignette to ground discussions on the military use of 5G during the SME workshop. This vignette consists of the following phases of operations before a conflict: reception, staging, onward movement, and integration (RSOI).[17] The vignette consists of four parts as follows:

---

[17] RSOI phases are commonly used in Army doctrine and operations literature. See, for example, A. Martin Lidy, Douglas P. Baird, John M. Cook, Robert C. Holcomb, Samuel H. Packer, and William J. Sheleski, *Doctrine, Organizations, and Systems for Reception, Staging, Onward Movement, and Integration (RSOI) Operations*, Alexandria, Va.: Institute for Defense Analyses, 1997, and Mark T. Simerly, *Improving Reception, Staging, Onward Movement, and Integration Operations for the Interim and Objective Forces*, thesis, Fort Leavenworth, Kan.: U.S. Army Command and General Staff College, 2002.

- *Reception* refers to the phase during which personnel and equipment arriving by sea and air are received at designated port and airfield locations. This includes, for example, offloading vessels, staging equipment and containers, uploading equipment to road and rail infrastructure, and offloading of aircraft.
- *Staging operations* occur when deployed units conduct a linkup; equipment is reassembled; units conduct theater-specific training; and units are scheduled to be moved to tactical assembly areas (TAAs) for integration. This includes such tasks as establishing intermediate staging bases (ISBs).
- Staging is then followed by *onward movement* when units are ready to move to TAAs. This can occur more than once depending on the size of the theater and the distance to and from ports (sea or air) of debarkation to TAAs. The onward movement phase can be the most challenging and time-consuming phase. Tasks include, but are not limited to, coordinating with multinational forces; reloading and preparing units for movement from ISBs to TAAs; and providing security or regional force protection.
- The last phase of the vignette is *integration*, when deployed units move to the tactical theater location, close the TAA, and begin battle handoff between outgoing and incoming units. This includes, among other tasks, transitioning to begin combat operations; communicating across services and task forces; configuring for combat; transitioning to defensive and offensive operations; conducting arming, fueling, and fixing equipment during combat; and regenerating the force during combat losses.

Organization of This Report

In Chapter 2, we discuss the rollout of 5G in the Baltics and other countries of interest. In Chapter 3, we describe Russian perspectives on military uses of 5G. Chapter 4 describes past and theorized attacks on cellular networks in Europe and Russian cellular jammers. In Chapter 5, we discuss potential benefits of using 5G in a smart logistics vignette. Finally, Chapter 6 concludes with recommendations.

Chapter 2. Expected 5G Rollout in Europe

In December 2021, the share of the population in Europe covered by at least fourth generation long-term evolution (4G LTE) or Worldwide Interoperability for Microwave Access was 100 percent in urban areas and 94 percent in rural areas (98.7 percent total).[18] The rollout of 5G, however, differs by country and regions within countries, and some locations are farther ahead than others.

The European Commission's Directorate-General for Communications Networks, Content and Technology is responsible for making Europe "fit for the digital age."[19] This is accomplished by using funding, legislation, and policy, and 5G implementation falls within this mandate. The European Commission created key organizations and policies for 5G, including the 5G Infrastructure Public Private Partnership (5G PPP) in 2013,[20] the European 5G Action Plan in 2016,[21] and the European 5G Observatory in 2018.[22] In 2021, the European Commission also published *2030 Digital Compass Communication: The European Way for the Digital Decade*, which lays out a vision for the European Union (EU)'s trajectory of digital technology.[23]

The 5G PPP is a partnership between the European Commission and the 5G Infrastructure Association. It is composed of the European information and communication technology (ICT) industry (ICT manufacturers, telecommunications operators, service providers, small and medium enterprises, and research institutions),[24] and it consists of three project phases: Phase 1, research; Phase 2, optimization; and Phase 3, large-scale trials.[25] During Phase 1, the 5G PPP defined 5G for EU purposes as mobile technology possessing the following characteristics when compared with 2010:

---

[18] S. O'Dea, "Estimated Share of Population Covered by at Least LTE or WiMAX Mobile (Cellular) Network Worldwide and in Rural and Urban Areas in 2021, by Region," Statista, December 8, 2021.

[19] European Commission, "What We Do—Communications Networks, Content and Technology: Mission Statement of the Directorate-General for Communications Networks, Content and Technology (Connect)," webpage, undated-m.

[20] 5G PPP, "Development of the 5G Infrastructure PPP in Horizon 2020," webpage, undated-c.

[21] European Commission, "5G Action Plan," webpage, last updated February 24, 2022c; and Margarita Robles-Carrillo, "European Union Policy on 5G: Context, Scope and Limits," *Telecommunications Policy*, Vol. 45, No. 8, 2021.

[22] European 5G Observatory, "What Is the European 5G Observatory?" webpage, undated.

[23] European Commission, *Communication from the Commission to the European Parliament, the Council, the European Economic and Social Committee and the Committee of the Regions, 2030 Digital Compass: The European Way for the Digital Decade*, Brussels, March 9, 2021.

[24] 5G PPP, "About the 5G PPP," webpage, undated-b.

[25] European Commission, *The 5G Infrastructure Public Private Partnership (5G PPP): First Wave of Research & Innovation Projects*, Brussels, undated-a.

1000 times higher mobile data volume per area, 10–100 times higher typical user data rate, 10 to 100 times higher number of connected devices, 10 times longer battery life for low-power devices, and five times reduced end-to-end (E2E) latency, reaching a target of 5 ms for road safety applications.[26]

The European Commission published its 5G Action Plan in 2016 outlining eight action items, priorities, and milestones to deploy 5G infrastructure for EU member states.[27] Overall, the EU 5G policy objectives include providing 5G coverage in all populated areas by 2030, deploying 5G corridors between member states, improving 5G network security, ensuring competition among 5G vendors and avoiding reliance on any single source, and stimulating EU 5G equipment manufacturing abilities.[28] European operators with user rights in a frequency band can decide which technologies they want to use in that band; however, the European Commission identified three pioneer frequency bands for 5G use so that member states can set common technical conditions and companies can manufacture 5G equipment that operate at the same frequencies across the EU.[29] The three EU pioneer bands are low-band (700 MHz), mid-band (3.6 GHz [3.4 to 3.8 GHz]), and high-band (26 GHz [24.25 to 27.5 GHz]).[30] The 5G Action Plan set a common date for releasing pioneer frequency bands for commercial use through license auctions, introducing limited commercial 5G services by the end of 2018, and starting large-scale deployment by the end of 2020. A spectrum license gives the license holder the rights to use their assigned spectrum for a particular purpose within a specified geographic area. The 5G rollout plan followed a general outline for member states, including a national strategy, public 5G spectrum consultations, spectrum auctions within each of these pioneer bands, and 5G projects.

To track the progress of member states toward the EU's goals and implementation of the 5G Action Plan, the European Commission established the 5G Observatory in 2018.[31] The 5G Observatory also analyzes 5G market trends across the EU and major international markets, technological progress, vertical industries, and use cases and publishes a quarterly report

---

[26] Jose F. Monserrat, Genevieve Mange, Volker Braun, Hugo Tullberg, Gerd Zimmermann, and Ömer Bulakci, "METIS Research Advances Towards the 5G Mobile and Wireless System Definition," *EURASIP Journal on Wireless Communications and Networking*, No. 53, 2015.

[27] European Commission, *Communication from the Commission to the European Parliament, the Council, the European Economic and Social Committee and the Committee of the Regions—5G for Europe: An Action Plan*, Brussels, September 14, 2016.

[28] European 5G Observatory, undated.

[29] European Commission, "European Commission to Harmonise the Last Pioneer Frequency Band Needed for 5G Deployment," press release, Brussels, May 14, 2019; and Belgian Institute for Postal Services and Telecommunications, "5G," webpage, undated-a.

[30] European 5G Observatory, 2021d.

[31] Robles-Carrillo, 2021.

detailing the progress of member states and these trends.[32] Most member states faced setbacks in their 5G rollouts because of the coronavirus disease 2019 (COVID-19) pandemic.[33]

Realizing the increasing strategic importance and change in the role and perception of digitalization during the COVID-19 pandemic, the European Commission released a new approach for a digitalized economy and society in its report, *2030 Digital Compass: The European Way for the Digital Decade*, which emphasized the 5G connectivity.[34] The *2030 Digital Compass* emphasizes investment and policies in 5G to enable a secure and sustainable digital infrastructure with the goal of ensuring a gigabyte network covers every European household and 5G covers all populated areas by 2030.[35] The European Investment Bank goes further, stating that the EU's 5G rollout will be "one of the most critical building blocks of the European digital economy and society in the next decade."[36]

The EU emphasizes industry verticals that will benefit from 5G, especially the automotive industry and railways and public transport. Connected and automated mobility (CAM) is a flagship use case for the EU's 5G strategy. The purpose of CAM is to create "complete ecosystems around vehicles, from road safety or digital rail operations to high-value commercial services for road users and train passengers."[37] Twenty-nine countries agreed to designate 5G cross-border corridors that allow tests and demonstrations of CAM technologies and vehicles to cross borders.[38] These cross-border tests involve 11 country borders. The trial projects that include our countries of interest are as follows:

- 5G-Mobix, which will involve 5G trials on urban corridors in Germany, the Netherlands, and Finland; the purpose is to test functionality of autonomous driving under various conditions[39]
- 5G-ROUTES, which will conduct CAM trials between Estonia, Finland, and Latvia over motor, rail, and ship transit routes; these large-scale trials will be done under "realistic conditions" across the Via Baltica-North corridor to validate 5G capabilities and 3GPP specifications[40]

---

[32] European Commission, "5G Observatory," webpage, last updated February 24, 2022d.

[33] Robles-Carrillo, 2021.

[34] European Commission, 2021.

[35] European Commission, 2021.

[36] François Gilles and Jaroslav Toth, *Accelerating the 5G Transition in Europe: How to Boost Investments in Transformative 5G Solutions*, Brussels: European Commission, February 2021, p. 5.

[37] Frédéric Pujol, Carole Manero, Basile Carle, and Santiago Remis, *5G Observatory Quarterly Report 12: Up to June 2021*, Brussels: European Commission, July 2021.

[38] European Commission, "Connected and Automated Mobility," webpage, last updated February 24, 2022e.

[39] European Commission, "5G Cross-Border Corridors," webpage, last updated February 22, 2022b; 5G PPP, "5G-MOBIX," webpage, undated-a.

[40] European Commission, 2022b; 5G-ROUTES, homepage, undated.

- 5G-Blueprint, which will conduct CAM trials between Belgium and the Netherlands over motor and ship transit routes and in ports; these trials will test and validate "uninterrupted cross-border teleoperated transport based on 5G connectivity"[41]
- 5GRAIL, which will test a future rail mobile communication system across various sites in Europe.[42]

In addition to the 5G Observatory, the European Commission monitors member states' digital progress through the Digital Economy and Society Index (DESI). With respect to 5G, the EU measures 5G readiness and 5G coverage as two indicators within connectivity. The European Commission assesses 5G readiness by the amount of spectrum within the pioneer bands assigned by each member state and readiness for 5G and 5G coverage by the percentage of populated areas covered by 5G.[43]

This chapter describes 5G activity through February 2022 in Estonia, Latvia, Lithuania, Belgium, Finland, Germany, the Netherlands, Norway, Poland, and Sweden. These countries were chosen as relevant to a Baltic states military scenario involving both the forward tactical areas and the support rear areas required to move equipment and supplies into theater. We provide a brief description of each country's 5G status and challenges in deployment. In Appendix B, we provide a summary of each country's national strategy or equivalent, when available; status of spectrum license auctions; and carriers and network coverage.

Estonia

As of this writing, Estonia lags behind the rest of Europe in deploying commercial 5G services because of spectrum auctions. In the *DESI 2021*, Estonia scored 0 percent in both 5G readiness and 5G coverage.[44] Estonian telecommunications companies still face spectrum auction delays but are otherwise prepared for 5G deployment. The telecommunications companies have selected hardware vendors and are also subsidiaries of companies that have successfully deployed 5G in other European countries.

Latvia

According to *DESI 2021*, Latvia was among the 5G frontrunners based on its relatively early 3.4 to 3.8 GHz spectrum auction but stagnated through 2020. Of particular concern was Latvia's

---

[41] European Commission, 2022b; 5G-Blueprint, homepage, undated.

[42] European Commission, 2022b; 5GRAIL, homepage, undated.

[43] European Commission, *Digital Economy and Society Index (DESI) 2021: DESI Methodological Note*, Brussels, undated-c, p. 6.

[44] European Commission, *Digital Economy and Society Index (DESI) 2021: Estonia*, Brussels, undated-d.

delayed 700 MHz spectrum auction.[45] Now that Latvia held its spectrum auctions, its telecommunications companies can begin rolling out 5G.

Lithuania

Lithuania faces challenges with neighboring countries Belarus and Russia using the pioneer frequencies. Belarus is using the 700 MHz band for broadcasting services and Russia is using the 3.5 GHz band for fixed and satellite services.[46] Russia confirmed that it plans to continue using the 3.5 GHz band for those purposes in August 2020, and the two countries have not come to an agreement on coordination. As a result of these coordination issues, Lithuania has been forced to delay its 5G spectrum auctions and is behind other EU countries. If these issues go unresolved, 5G will be blocked in a major part of the country.[47]

Belgium

Other than delayed spectrum auctions, Belgium faces challenges on radiation limits and differences between its three regions. Each region has different radiation limit standards and imposes costs that DESI reports state do not "allow economically viable 5G deployment,"[48] particularly in the Brussels-Capital Region. The regional governments in Flanders and Brussels plan to change radiation limits to make 5G deployment more attractive. Additionally, permits allowing companies to deploy antennas and fees for pylons can reach 10,000 euros per year for each antenna,[49] making 5G deployment particularly expensive because it requires more antennas than previous generations.

Finland

Finland also faces challenges in the 3.5 GHz frequency band based on coordination issues with Russia. According to the Finnish Ministry of Transport and Communications, the lack of an agreement with Russia will place considerable restrictions on Finland, possibly affecting several hundred kilometers from the border for the 3.6 to 3.8 GHz range and 100 km for the 3.4 to 3.6

---

[45] European Commission, *Digital Economy and Society Index (DESI) 2021: Latvia*, Brussels, undated-g, p. 11.

[46] European Commission, *Digital Economy and Society Index (DESI) 2021: Lithuania*, Brussels, undated-h, p. 9.

[47] European Commission, undated-h, p. 9.

[48] European Commission, *Digital Economy and Society Index (DESI) 2021: Belgium*, Brussels, undated-b, p. 9.

[49] European Commission, undated-b, p. 9.

GHz range.[50] Another challenge is a lack of incentives for private investment in Finland's sparsely populated areas and its reliance on public funding to reach its coverage goals.[51]

Germany

Deutsche Telekom, Telefónica Germany, and Vodafone Germany each launched 5G services in Germany. Deutsch Telekom announced that it began rolling out 5G base station antennas in May 2018 and began 5G services in the 3,700 MHz band using a test license. After receiving a license during the 3.5 GHz spectrum auction, Deutsche Telekom transitioned to its full commercial spectrum license in September 2019.[52] Vodafone Germany began deploying its narrow band IoT network in March 2018 and had deployed 90 percent of its network by October 2018.[53] Vodafone started its commercial 5G network in July 2019. By May 2021, Vodafone's network covered 25 million people and Vodafone Germany planned to cover 30 million people by the end of 2021, which was an increase from its original goal of 20 million.[54] Telefónica Germany introduced commercial 5G services in October 2020, and its network operates in the 3.5 GHz spectrum band.[55] In July 2021, Telefónica's network covered 80 cities in Germany.[56]

The Netherlands

The Netherlands has been advancing its 5G deployment and coverage since 2020, but it has only auctioned the 700 MHz band so far.[57] Its overall 5G readiness is at 33 percent and behind the EU average.[58] This is largely due to the delays in the 3.5 GHz spectrum auction, and 5G rollout will continue to lag until the Netherlands resolves legal issues with Inmarsat.[59]

---

[50] Ministry of Transport and Communications, *Turning Finland into the World Leader in Communications Networks—Digital Infrastructure Strategy 2025*, Helsinki, 2019.

[51] European Commission, *Digital Economy and Society Index (DESI) 2021: Finland*, Brussels, undated-e, p. 11.

[52] HBR Radiofrequency Technologies, "Telekom Deutschland," webpage, undated-d.

[53] HBR Radiofrequency Technologies, "Vodafone Germany," webpage, undated-e.

[54] Juan Pedro Tomás, "Vodafone Germany Expands 5G Footprint, Adds New 5G Devices," RCR Wireless, August 13, 2021b.

[55] Pujol et al., 2021.

[56] Juan Pedro Tomás, "Telefonica Deutschland Reaches 80 German Cities with 5G Technology," RCR Wireless, July 6, 2021a.

[57] "KPN, T-Mobile and VodafoneZiggo Acquire Frequencies in Dutch Mobile Communications Auction," Government of the Netherlands, July 21, 2020.

[58] European Commission, *Digital Economy and Society Index (DESI) 2021: Netherlands*, Brussels, undated-i, p. 7.

[59] "Inmarsat Wins Injunction Against Dutch Refarming of 3.5 GHz Band," Telecompaper, June 30, 2021.

Norway

In March 2020, Telenor Norway became the first carrier to offer commercial 5G services. Norway is divided into 11 counties, and as of this writing, Telenor offers 5G services in at least one city for ten of these counties.[60] Telenor Norway plans to expand coverage into all 11 counties in 2022 and provide a nationwide network by 2024. Telia Norway activated its 5G network in May 2020, initially covering Lillestrøm and parts of Oslo; it planned to cover half of the Norwegian population by the end of 2020 and the entire country by 2023.[61] By September 2021, Telia Norway had covered approximately 30 percent of the population and had a goal of reaching half of the population by the end of 2021.[62] Ice started offering 5G services in September 2020. By November 2021, Ice's network had covered approximately 10 percent of Oslo's population. Ice plans to cover four of Norway's five largest cities and 75 percent of the population in the short term.[63]

Poland

One of the primary challenges to Poland's 5G rollout is the delays to its 700 MHz, 3.6 GHz, and 26 GHz spectrum auctions. As of January 2022, Poland had not assigned any of these frequencies for 5G deployment. According to DESI reports, "swift assignment will be necessary for the provision of 5G connectivity under transparent, open and nondiscriminatory conditions."[64] As with Lithuania, another challenge for Poland is deconflicting the 5G pioneer bands with Russia in Kaliningrad.[65] In December 2018, the Polish Senate approved the delay of the 700 MHz spectrum auction from June 2020 to June 2022 because of potential coordination issues with Russia.[66]

Sweden

The Ministry of Enterprise and Innovation identified that the foundation of its broadband and mobile strategy should be market driven expansion, but concerns about the profitability of

---

[60] "Telenor to Provide 5G in All 11 Norwegian Counties in 2022, Targeting 250 New Sites," Telecompaper, September 30, 2021.

[61] "Telia Launches 5G in Norway," Telia Company, May 12, 2020.

[62] "Telia Norge's 5G Network Now Covers 30% of the Population," CommsUpdate, September 22, 2021.

[63] Mary Lennighan, "Norway to Get More Competitive as Ice Steps Up," Telecoms, November 18, 2021.

[64] European Commission, *Digital Economy and Society Index (DESI) 2021: Poland*, Brussels, undated-k, p. 3.

[65] John C. K. Daly, "Lithuanian-Russian Radio Frequency Dispute Highlights Problems of Civilian Versus Military Applications," *Eurasia Daily Monitor*, Vol. 18, No. 38, March 8, 2021.

[66] "Polish Senate Approves Delay to 700MHz Switchover," CommsUpdate, March 1, 2019.

private investment in Sweden's rural and sparsely populated areas exist.[67] 5G is an important component of Sweden's broadband strategy.[68] According to the DESI country profile, "the successful deployment of 5G in Sweden depends on the timely availability and assignment of the 5G pioneer bands."[69] Despite earlier setbacks and delays in the 3.5 GHz spectrum auction, Sweden completed allocation of the auction.

Summary

Most countries considered here are behind the rest of the EU in 5G DESI scores of readiness and coverage; these scores are summarized in Table 2.1. In comparison with the average EU DESI score of 51 percent in 5G *readiness* (defined as the amount of spectrum within the pioneer bands assigned by each member state that are ready for 5G), only Finland and Germany are at or above average (99 percent and 100 percent, respectively). Norway and Sweden are not far off, both with readiness scores of 49 percent. Compared with an average EU DESI score of 14 percent in 5G *coverage* (defined as the percentage of populated areas covered by 5G), only Germany, the Netherlands, and Sweden score on par or higher, with 18 percent, 80 percent, and 14 percent coverage, respectively. The Netherlands' high 5G coverage score reflects its goal to be the leader of 5G in Europe; the Netherlands' 700 MHz band spectrum licenses obligate license holders to cover 98 percent of the geographical area of municipalities throughout the country.[70] Finland and Poland have 5G coverage scores relatively close to the EU average at 12 percent and 10 percent, respectively.

---

[67] Government Offices of Sweden, *A Completely Connected Sweden by 2025—a Broadband Strategy*, Stockholm, undated, p. 7.

[68] European Commission, *Digital Economy and Society Index (DESI) 2021: Sweden*, Brussels, undated-l, p. 8.

[69] European Commission, undated-l, p. 8.

[70] European Commission, undated-i, p. 8.

Table 2.1. European Commission Digital Economy and Society Index 2021 5G Scores

Country	5G Readiness (%)	5G Coverage (%)
EU average	51	14
Estonia	0	0
Latvia	29	0
Lithuania	5	0
Belgium	3	4
Finland	99	12
Germany	100	18
The Netherlands	33	80
Norway	49	5
Poland	0	10
Sweden	49	14

SOURCES: European Commission, undated-d; European Commission, undated-g; European Commission, undated-h; European Commission, undated-b; European Commission, undated-e; European Commission, *Digital Economy and Society Index (DESI) 2021: Germany*, Brussels, undated-f; European Commission, undated-l; European Commission, *Digital Economy and Society Index (DESI) 2021: Norway*, Brussels, undated-j; European Commission, undated-k; and European Commission, undated-l.

All countries of interest are in the process of deploying 5G, and each is at a different point in the process. Table 2.2 provides a summary of each country's 5G operators, commercial launch dates, and frequency bands. Coverage within each country varies, generally with far greater coverage over urban areas and major corridors than rural areas. Many delays are due to spectrum conflict issues in the countries that share borders with Russia, which complicates their 5G rollout. However, all countries target a full deployment date between 2025 and 2030.

Table 2.2. Summary of Commercial 5G Rollout in Countries of Interest

Country	Operator	5G Launch Date	Band
Belgium	Proximus	April 2020	2.1 GHz
	Telenet Belgium	December 2021	3.6–3.8 GHz (temporary license)
	Orange Belgium	February 2022	3.6–3.8 GHz (temporary license)
Estonia	Telia Estonia	November 2020	Unclear (3.5 GHz auction pending)
	Tele2 Estonia	Expected 2022	N/A
	Elisa Estonia	Expected 2022	N/A
Finland	Elisa Finland	June 2018	3.6 GHz
	Telia Finland	November 2019	3.5 GHz
	DNA (owned by Telenor)	January 2020	3.5 GHz
Germany	Vodafone Germany	July 2019	800 MHz, 1.8 GHz, 3.6 GHz
	Deutsche Telekom	September 2019	2.1 GHz, 3.6 GHz
	Telefónica	October 2020	3.6 GHz
Latvia	Latvijas Mobilais Telefons (LMT)	July 2019	3.5 GHz
	Tele2 Latvia	January 2020	3.5 GHz
	Bite Latvia	January 2021	Unclear[a]
Lithuania	Telia Lithuania	November 2020	2.1 GHz, 3.5 GHz
	Bite Lithuania	Expected 2022	N/A
	Tele2 Lithuania	Expected 2022	N/A
The Netherlands	Vodafone Ziggo	April 2020	800 MHz, 1.8 GHz, 2.1 GHz, 2.6 GHz
	T-Mobile	July 2020	700 MHz
	Koninklijke PTT Nederland (KPN)	July 2020	700 MHz
Norway	Telenor Norway	March 2020	700 MHz
	Telia Norway	May 2020	3.6 GHz
	Ice	September 2020	700 MHz, 2.1 GHz, 3.6 GHz
Poland	Plus	May 2020	2.6 GHz
	T-Mobile Poland	June 2020	2.1 GHz
	Play	June 2020	2.1 GHz
	Orange Poland	July 2020	2.1 GHz
Sweden	Telia	May 2020	700 MHz
	Tele2	May 2020	3.6 GHz
	Tre Sweden	June 2020	3.4–3.5 GHz
	Telenor Sweden	October 2020	3.7 GHz

NOTE: N/A = not applicable. Table is up to date as of February 2022. Coverage within each country varies.
[a] We were unable to confirm the frequencies in use by this operator.

Chapter 3. The Russian Military and 5G

This chapter discusses Russian military views on 5G on the basis of open-source Russian-language publications, including both military specialist literature and trade newspapers. Like their Western counterparts, Russian defense thinkers are enthusiastic about the military potential of 5G technologies. Ironically, however, the primary obstacle between the Russian military and the possible benefits of 5G is posed by the Russian military itself. As of this writing, the 3.4 to 3.8 GHz band that serves as the backbone of the emerging global 5G standard is used by the Russian military for other purposes, and the Russian military has proved extremely resistant to calls to reallocate this spectrum for civilian applications. Instead, the Russian government has proposed an alternative 5G standard using the 4.8 GHz band.[71]

Russian commercial interests have pushed back against this proposal because it would prevent the use of standard equipment and could cripple 5G rollout to Russian consumers. But if Russia continues its flirtation with the alternative 4.8 GHz standard for commercial use, it might inconvenience more than Russian mobile subscribers: Because of international treaty obligations regarding spectrum use in neighboring countries, it would also constrain the freedom of NATO to use standard 5G equipment close to Russia's borders.[72] Furthermore, in the 2022 invasion of Ukraine, it became apparent that Russian ground forces were remarkably dependent on the cellular networks of occupied territory for their communications, suggesting that they might try to exploit Western 5G networks in a future conflict scenario.[73] Some Russian military communications equipment required civilian cellular networks to operate, while Russian commanders often relied on civilian phones because of the flakiness of the radios issued to them. As a consequence, the seemingly trivial issue of whether Russia ultimately rejects the 3.4 to 3.8 GHz 5G standard matters considerably for the role of 5G in a NATO-Russia confrontation.

If 5G plays a role in a future conflict between Russia and NATO, it will do so in the context of Russia's homegrown concept of *radioelectronic struggle (radioelektronnaia bor'ba* [REB]). Irrespective of whether 5G is used by the Russian military, its NATO adversaries, or both, this Russian counterpart to EW will attempt to shape and constrain what 5G can and cannot accomplish on the battlefield. As Russian analysts themselves regularly point out, despite considerable parallels, Russian REB differs qualitatively from U.S. concepts of EW, most

---

[71] David George, Dennisa Nichiforov-Chuang, and Emanuel Kolta, *How Spectrum Will Shape the Outlook for 5G in Russia*, London: GSMA, 2020.

[72] "Lithuania's 5G Development Hampered by Russian Military Infrastructure," LRT English, February 23, 2021.

[73] Sergei Dobrynin and Mark Krutov, "Communication Breakdown: How Russia's Invasion of Ukraine Bogged Down," RFERL, March 19, 2022.

fundamentally in that it pursues different goals.[74] REB is conceptually distinct from other staples of Russian military thought, such as *maskirovka* (roughly similar to camouflage, concealment, and deception [CCD]) and information warfare, but intimately connected with them in practice.

Russian Perspectives on Electronic Warfare

Russia's distinctive approach to EW developed over the course of many decades. The Russian military was a pioneer in radio applications—some of the very earliest experiments in wireless telegraphy were carried out under the auspices of the Imperial Russian Navy—but its thinking and planning for what we would today call EW remained limited until the 1950s. Up until that time, the Soviet military employed a concept known in Russian as *radioprotivodeistvie* or *radio countermeasures*. These included the components of EW practiced during the Second World War, such as radar jammers. But during the early stages of the Cold War, the Soviet Union's capitalist adversaries began introducing a wide array of qualitatively new electronic countermeasures, challenging the Soviet military both to develop means to neutralize these electronic countermeasures and to counterbalance them with similar equipment of its own. This required not just the cultivation of a new technological base but also new modes of thinking. The individual most associated in Russia with these revolutions was Viktor Ivanovich Kuznetsov (1920–2016), a Soviet general whose involvement in EW began when he served as a radar engineer during the Great Patriotic War. In the late 1950s, Kuznetsov was tasked with developing jammers against adversary aircraft radars, which in turn evolved into an effort to conceptualize a comprehensive approach to conflict involving the EMS. Kuznetsov championed a proposal to establish a dedicated institute to study EW.[75] This led to the foundation of the Ministry of Defence interservice research and testing center on EW problems (the 21st Scientific-Technical Testing Center of the Russian Ministry of Defense, or *21 NIITs MO*) in Voronezh, which Kuznetsov headed until 1986.[76]

A Framework for Thinking about Russian Radioelectronic Struggle

In Kuznetsov's analysis, what later became known as REB was defined as "as a system of organizational and technical measures to disorganize the control of enemy forces and weapons

---

[74] I. A. Lastochkin, I. E. Donskov, and A. L. Morstaresku, "An Analysis of Contemporary Conceptions for Undertaking Operations in the Electromagnetic Spectrum from the Standpoint of Radioelectronic Struggle" ["Analiz covremmennykh kontseptsii po vedeniiu operatsii v elektromagnitnom spektre s pozitsii radioelektronnoi bor'by"], *Military Thought* [*Voennaia Mysl'*], No. 4, 2021, pp. 36–37.

[75] V. I. Kharpukhin, "On the 100th Anniversary of the Birth of the Formidable Scholar in the Theory and Practice of Electronic Warfare V. I. Kuznetsov" ["K 100-letiiu so dnia rozhdeniia krupnogo uchenogo v oblasti teorii i praktiki radioelektronnoi bor'by V. I. Kuznetsova"], *Military Thought* [*Voennaia Mysl'*], No. 12, 2020, p. 109.

[76] Kharpukhin, 2020, p. 114.

and ensure the stable functioning of the [Soviet Armed Forces'] own control systems."[77] The term *REB* emerged in the late 1960s and is still defined in terms similar to those Kuznetsov gave over six decades ago. A 2017 article by Russian EW officers defined it as follows:

> REB is a set of coordinated activities and actions encompassing radioelectronic attack on adversarial radioelectronic and information-technical assets, radioelectronic protection of radioelectronic and information-technical assets, countermeasures against technical reconnaissance and radioelectronic information support measures.[78]

REB differs from the earlier concept of *radio countermeasures*; although this older formulation was purely defensive and reactive, REB encompasses both offensive and defensive aspects.[79] These are dubbed *radioelectronic destruction* and *radioelectronic protection*, respectively. Since 2000, Russian thinkers have fleshed out the REB framework further to add a third aspect, *radioelectronic-informational support*. In turn, each of these three aspects encompasses multiple subcategories (see Figure 3.1).

Radioelectronic Destruction

Radioelectronic destruction has three subcategories. The first is *radioelectronic suppression*, which ironically includes nonradioelectronic aspects. Much of its focus consists of jamming or suppressing the adversary's ability to employ the EMS or analogous means, such as hydroacoustics, against the interests of the Russian military. Along with suppression of radio, electro-optical, acoustic, and hydroacoustic means, radioelectronic suppression involves modifying the conditions of electromagnetic (EM) transmission and reflection.[80] The second subcategory, *defeat by guided anti-radiation munition*, focuses on countermeasures against missiles with passive guidance systems that seek out specific EM, optical, or acoustic signatures.[81] The third subcategory, *functional radioelectronic destruction*, involves "destruction by electromagnetic emissions" and "destruction by special programmed means." The latter is a way "to reduce the functional efficiency or disable components of the information processing systems of enemy radioelectronic means," and as "violation of the confidentiality, integrity and availability of information through the use of special software" (i.e., cyberattacks).[82]

---

[77] Kharpukhin, 2020, p. 109.

[78] V. F. Guzenko and A. L. Moraresku, "Radioelectronic Struggle: Contemporary Substance" ["Radioelektronnaia bor'ba: Sovremennoe soderzhanie"], Radioelectronic Struggle in the Armed Forces of the Russian Federation [*Radioelektronnaia bor'ba v Vooruzhennykh Silakh Rossiiskoi Federatsii*], 2017, pp. 14–16.

[79] Kharpukhin, 2020, p. 109; and S. I. Makarenko, *Information Confrontation and Radioelectronic Struggle in the Network-Centric Wars of the Early 21st Century* [*Informatsionnoe protivoborstvo i radioelektronnaia bor'ba v setetsentricheskikh voinakh nachala XXI veka*], Saint Petersburg, Russia: Naukoemkie Tekhnologii, 2017, p. 81.

[80] Makarenko, 2017, p. 82.

[81] Makarenko, 2017.

[82] Makarenko, 2017, p. 83.

Figure 3.1. Classification of Radioelectronic Struggle Measures

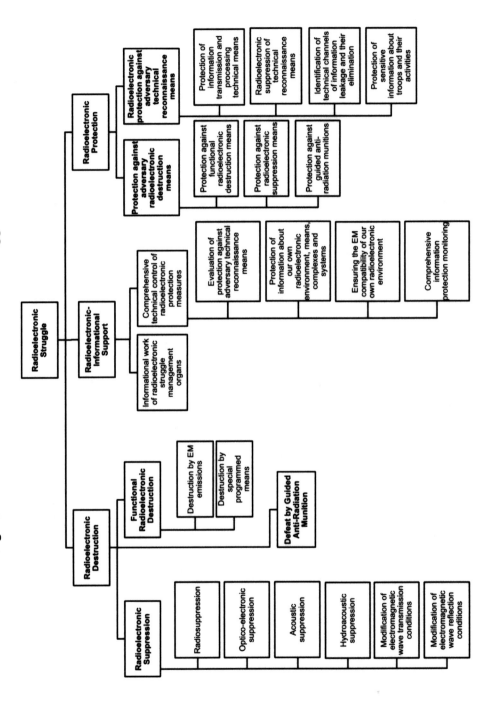

Radioelectronic Protection

Radioelectronic protection encompasses two major areas of activity. The first of these, "protection against adversary radioelectronic destruction means," is basically a mirror image of the *radioelectronic destruction* category. It consists of protective measures against "radioelectronic suppression," "functional radioelectronic destruction," and adversary antiradiation and other passive-guided missiles.[83]

The other subcategory, "radioelectronic protection against adversary technical reconnaissance means," is broad but fairly self-explanatory. Along with radioelectronic suppression of adversary reconnaissance assets, it consists of the protection of the technical means of information transmission and processing, the protection of information about troops and their activities, and the identification of the sources of information leaks and their remediation.[84] As these latter activities suggest, there exists considerable overlap between REB and Russian notions of *informatsionnoe protivoborstvo* (roughly, *information warfare*) and *maskirovka* (analogous to CCD). The perceived importance of these activities is illustrated by the fact that some Russian military writers classify "radioelectronic defense against adversary technical reconnaissance means" as a fourth major aspect of REB.[85]

Radioelectronic Informational Support

The third major aspect of REB is *radioelectronic informational support*, which is defined as "a set of measures and actions to identify the functioning of the enemy's radioelectronic means to enable their electronic destruction and to control the functioning of one's own radioelectronic means in order to protect them."[86] Radioelectronic informational support comprises two subactivities.

The first of these is "comprehensive technical control of radioelectronic defense measures," which seeks to ascertain the functioning of these systems in the face of adversary intelligence, surveillance, and reconnaissance (ISR) and EW, protect information about them, and keep track of the state of one's own EM environment.[87]

The second subactivity, "comprehensive control of information defense measures," aims to identify target indicators that might be revealed by friendly forces during combat employment so that they can be suppressed, along with "the prompt suppression of violations of norms and regulations regarding counterintelligence."[88] The second subactivity of radioelectronic

---

[83] Guzenko and Mor"resku, 2017.

[84] Makarenko, 2017, p. 83.

[85] Makarenko, 2017.

[86] Makarenko, 2017, p. 83.

[87] Makarenko, 2017, p. 84.

[88] Makarenko, 2017, p. 84.

informational support, the "informational work of radioelectronic struggle management organs," "consists of collecting, accumulating, analyzing, synthesizing, storing and distributing data on adversary and friendly radioelectronic means" and achieving "the complex technical control of electronic protection measures."[89] In addition to signals intelligence, exploited data for these purposes can be acquired via "espionage, military aircraft accidents, the export of military equipment, and, in wartime, the capture of enemy radioelectronic equipment."[90]

Russian Versus U.S. Concepts of Electronic Warfare

As Russian military writers themselves regularly emphasize, this framework differs in several fundamental ways from Western and U.S. concepts of EW even though significant parallels exist. The differences between U.S. and Russian EW concepts were the subject of several articles in the prestigious journal of the Russian General Staff, *Voennaia Mysl'* (*Military Thought*) published since 2014. The authors of one of these articles drew particular attention to the absence of a Russian equivalent of such documents as the 2013 DoD *Electromagnetic Spectrum Strategy* or the 2020 *Electromagnetic Spectrum Superiority Strategy*.[91] In fact, despite the ubiquity of REB in Russian military thinking and operations, its presence in Russian strategy and doctrinal documents as of 2021 is relatively muted. For instance, REB does not even receive a dedicated paragraph in the 2014 version of the *Military Doctrine of the Russian Federation*, and it is only mentioned a handful of times in that document.[92] In fact, the official Russian government definition of *REB* was not promulgated by the Ministry of Defence at all but rather by the Commonwealth of Independent States standards agency, Euro-Asian Council for Standardization, Metrology and Certification, which is essentially the post-Soviet equivalent of the American National Standards Institute.[93]

According to the authors of the *Voennaia Mysl'* article, Russian strategists see their approach to REB as

> one of the effective, asymmetric and quickly implemented ways to counter the advantages of leading foreign states in the creation and use of electronic and information-control systems, that is, it is designed to neutralize their "superiority in the electromagnetic environment."[94]

---

[89] Makarenko, 2017, p. 84.

[90] Makarenko, 2017, p. 84.

[91] DoD, *Electromagnetic Spectrum Strategy*, Washington, D.C., September 2013; and DoD, *Electromagnetic Spectrum Superiority Strategy*, Washington, D.C., October 2020.

[92] President of Russia, *Military Doctrine of the Russian Federation* [*Voennaia doktrina Rossiiskoi Federatsii*], Moscow, December 26, 2014.

[93] Guzenko and Moraresku, 2017.

[94] Lastochkin, Donskov, and Moraresku, 2021, p. 36.

They contend, "Our forces and means of REB are employed directly and integrated into operations with other forces with a more significant goal in the art of war—to deprive the enemy of the opportunity to carry out combat missions in an organized and effective manner with his troops and weapons." They acknowledge, however, that the "role of the electronic warfare troops and their equipment in achieving this goal is constantly growing and becoming a leading one."[95]

As the authors intimate, the relative position of EW in both Russian military thought and institutional priorities has improved considerably over the past fifteen years, even though in some ways it still lacks the prominence that EW has in U.S. military doctrine. Russia's underwhelming military performance in the 2008 Russo-Georgian War led to the so-called Serdiukov reforms, one component of which was bolstering EW troops and introducing modernized EW equipment. In 2009, the Russian military introduced dedicated EW troops, which in the context of Russian military operations theory signaled that EW had been elevated from a combat support element to a combat support arm.[96] But the new, dedicated EW units represent only a modest fraction of the Russian military's EW personnel and equipment: Most of these remain integrated with other types of units, in keeping with Russia's theoretical approach to EW.[97]

Perhaps more important for Russia's EW capabilities than the dedicated EW troops was the long-overdue investment in modernized EW equipment. (See Chapter 4 of this report for a discussion of some Russian EW systems and their characteristics.) Prior to the Serdiukov reforms, Russian EW equipment dated back to the Soviet period and was often ill-maintained, while a lack of domestic sales left Russian producers of such equipment dependent on export orders for their survival. Since 2008, the Russian military has procured impressive new EW systems, such as the Krasukha-4 and Leer-3, and some of these systems have proved themselves in combat use in Syria. But despite these investments, a large fraction (possibly most) of EW equipment in use by the Russian military today still consists of legacy systems inherited from the Soviet Union.[98]

Another important difference between the U.S. approach to EW and its Russian counterpart is that even though various parts of the Russian armed forces are supposed to coordinate their actions in the EMS, there is no Russian equivalent to the kind of "joint electromagnetic spectrum operation" contemplated by Joint Publication 3-85.[99] In part, this reflects one of the basic

---

[95] Lastochkin, Donskov, and Moraresku, 2021, p. 37.

[96] Jonas Kjellén, *Russian Electronic Warfare: The Role of Electronic Warfare in the Russian Armed Forces*, Stockholm: Swedish Defence Research Agency, FOI-R—14625—SE, September 2018, pp. 29–30.

[97] Kjellén, 2018, p. 61.

[98] Roger N. McDermott, *Russia's Electronic Warfare Capabilities to 2025: Challenging NATO in the Electromagnetic Spectrum*, Tallinn: International Centre for Defence and Security, 2017, p. B-1; and Kjellén, 2018, p. 15.

[99] Joint Publication 3-85, *Joint Electromagnetic Spectrum Operations*, Washington, D.C.: U.S. Joint Chiefs of Staff, May 22, 2020.

distinctions between Russian and U.S. arrangements for military planning. Historically, U.S. armed services have enjoyed a very large degree of freedom of action and have tended to plan operations independently of the other services. Russia, by contrast, has had a general staff on the Prussian model since Tsarist times to which the various armed services are subordinated. But the more fundamental reason that Russia lacks joint EW doctrine is because the EMS is not regarded as a distinct combat domain the way that it is in contemporary U.S. military thought:

> [I]n our doctrinal documents, the electromagnetic spectrum (environment) is not considered as a special environment (sphere) of military operations. The forces and means operating in this environment (communications, reconnaissance, electronic warfare, etc.) are used according to their own plans, are controlled independently by the relevant control bodies (for communications, intelligence and EW) and coordinate their actions at the level of interaction.[100]

The Russian hesitation to reconceptualize the EMS as a domain in its own right is not due to simple institutional conservatism: Since 2000, Russian security doctrine has cultivated an elaborate theory of "information security" that characterizes the "information space" as a critical arena of international security competition.[101]

The authors of the *Voennaia Mysl'* article conclude on the basis of their analysis that

> it should be stated that American and our own approaches to conflict in the electromagnetic spectrum (environment) are somewhat similar and consonant, but at the same time their goals fundamentally differ. Whose approach is better is a philosophical question. There is confidence that our approach is adequate to the need and no worse [than the American one].[102]

Some Russians apparently believe that the U.S. approach to EW could be superior because articles advocating a shift toward a conceptual framework more similar to the American one have appeared in the Russian defense press. For instance, a 2015 *Voennaia Mysl'* article made the case that the EMS deserved to be elevated to the level of a full-fledged combat domain on par with land, sea, air, and space.[103] The following year, an article appeared in the same journal by different authors; they advocated that EW merited not just its own units, but its own dedicated combat arm, given the increasing importance of EW.[104] At present, however, no indicators have

---

[100] Joint Publication 3-85, 2020, p. 35.

[101] Ministry of Defence of the Russian Federation, *Conceptual Views on the Responsibilities of the Armed Forces of the Russian Federation in the Information Space* [*Kontseptual'nye vzgliady na deiatel'nost' Vooruzhennykh Sil Rossiiskoi Federatsii v informatsionnom prostranstve*], Moscow, 2011.

[102] Lastochkin, Donskov, and Morаresku, 2021, p. 38.

[103] I. A. Lastochkin, "Role and Place of Electronic Warfare in Contemporary and Future Combat Actions" ["Rol' i mesto radioelektronnoy bor'by v sovremennykh i budushchikh boyevykh deystviyakh"], *Military Thought* [*Voennaia Mysl'*], No. 12, 2015.

[104] I. Korolyov, S. Kozlitin and O. Nikitin, "Problems of Determining Ways of Employing Forces and Means of Electronic Warfare" ["Problemy opredeleniya sposobov boevogo primeneniya sil i sredstv radioelektronnoy bor'by"], *Military Thought* [*Voennaia Mysl'*], No. 9, 2016.

yet appeared that Russian defense officials are seriously contemplating either of these dramatic steps.

The Prospective Role of Russian Radioelectronic Struggle in Information Warfare and Camouflage, Concealment, and Deception

As noted earlier, Russian theorists primarily see EW (REB) as an enabler of other goals rather than as an end unto itself. This is exemplified by the manner in which these diverse objectives are integral to the various subcomponents of REB outlined earlier. Perhaps the single most important of these is *information warfare (informatsionnaia voina)*, which a 2011 Ministry of Defence document defined as

> a confrontation between two or more states in the information space with the aim of causing damage to information systems, processes and resources, critical and other structures, undermining political, economic and social systems, massive psychological manipulation of the population to destabilize society and the state, as well as coercion of states to make decisions in the interests of the opposing side.[105]

Russian theorists regard the information space (informatsionnoe prostranstvo) as an important site of contestation with potential adversaries, unlike the EMS. The *information space* is defined as "the field of activity associated with the formation, creation, transformation, transmission, use, and storage of information," affecting "individual and public consciousness, information infrastructure and information itself."[106] Any use of radioelectric means to affect adversary situational awareness by interfering with their ISR assets or communications capabilities is both REB and information war simultaneously.

Another major sphere of activity that intersects with both REB and information warfare is maskirovka. While usually translated into English as "camouflage," maskirovka is much broader; as noted earlier, it is analogous to CCD in U.S. military doctrine. A Russian military dictionary defined *maskirovka* as

> a complex of interrelated organizational, operational-tactical and engineering-technical measures carried out in order to conceal troops (forces) and facilities from the enemy and mislead him about their presence, location, composition, state, as well as plans of command, actions and intentions of troops (forces), preserving their combat ability and increasing the survivability of facilities.[107]

Three of the four types of maskirovka listed in the definition—radioelectronic, optical-electronic, and acoustic—intersect with the purview of REB as described earlier. In addition to traditional camouflage techniques, such as painting concealing patterns on equipment,

---

[105] Ministry of Defence of the Russian Federation, 2011.

[106] Ministry of Defence of the Russian Federation, 2011.

[107] "Camouflage" ["Maskirovka"], *Strategic Rocket Forces Encyclopedia* [*Entsiklopediia RVSN*], Ministry of Defense of the Russian Federation, undated.

maskirovka is also accomplished via such means as *disinformation* (*dezinformatsiya*), which "is carried out by spreading false information among military personnel and the local population in various ways."[108] According to these definitions, the use of spoofers transmitting fake signal traffic to obscure the location of a specific asset is simultaneously REB, information warfare, and maskirovka.

Russian military theorists believe that they must aggressively blend these various concepts to counter what they perceive as the threat from aggressive, technologically and militarily superior Western powers, particularly the United States. For decades, Russian analysts have expressed considerable alarm about American concepts of *network-centric warfare* and the kind of devastating "aerospace campaign" it might enable against Russia.[109] They fear that an ability to fuse information from ISR assets to track and target Russian military assets in real time, combined with deep inventories of precision conventional munitions, could enable the defeat of the Russian military in a swift nonnuclear operation. Logically, Russian analysts have concluded that disrupting anticipated U.S. capabilities to collect, transmit, and fuse data in real time is essential to avoid defeat in such a war, in which they anticipate the United States will engage in extremely aggressive EW measures of its own.[110]

A 2017 book, *Information Confrontation and Radioelectronic Struggle in the Network-Centric Wars of the Early 21st Century* (*Informatsionnoe protivoborstvo I radioelektronnaia bor'ba v setetsentricheskikh voinakh nachala XXI veka*) outlines several scenarios of the role EW might play in these "network-centric wars" against Russia. Figure 3.2 reproduces a timeline for one such conflict, illustrating how Russian analysts expect that EW to obfuscate the imminent assault would begin almost a full day before the launch of kinetic strikes.[111]

---

[108] "Camouflage," undated.

[109] Lastochkin, Donskov, and Moraresku, 2021, p. 30; and Thomas R. McCabe, "The Russian Perception of the NATO Aerospace Threat: Could It Lead to Preemption?" *Air & Space Power Journal*, Fall 2016.

[110] For an extended example of such an analysis, see Makarenko, 2017.

[111] Makarenko, 2017, p. 91.

Figure 3.2. A Russian Conception of Radioelectronic Struggle (Electronic Warfare) in the Opening Phase of a *Net-Centric War*

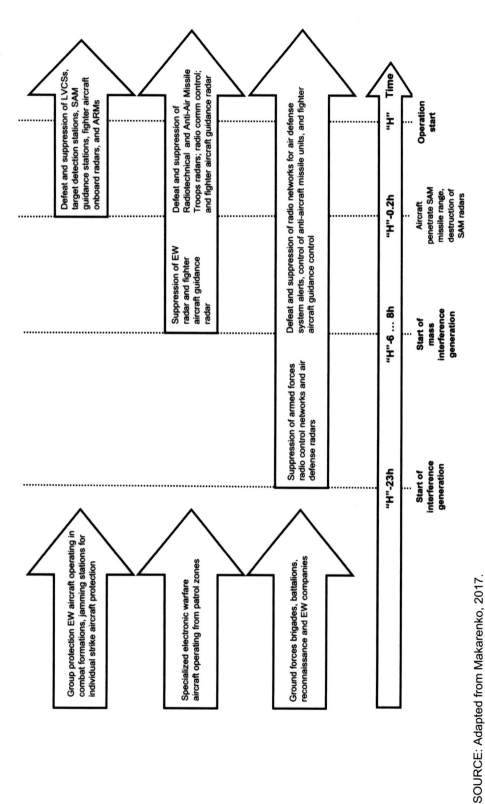

SOURCE: Adapted from Makarenko, 2017.
NOTE: ARM = antiradiation missile; comm = communications; LVCS = launch vehicle control system; SAM = surface-to-air missile. H-hour is the time at which kinetic operations start.

Russian Telecom Sector

Post-Soviet Russia inherited comparatively little salvageable civilian telecommunications infrastructure from the Soviet Union. As a consequence, post-Soviet telecommunications infrastructure evolved on a relative greenfield basis to emphasize mobile customers. As of this writing, Russia has very high customer penetration for mobile phone subscribers, which in turn is supported by extensive infrastructure for 4G and earlier standards. The Russian mobile service provider industry is highly competitive, and there is constant jockeying between what are known as the "big four" firms offering nationwide service: Mobile TeleSystems (MTS), Beeline, MegaFon, and Tele2.[112] All of these firms are Russian-controlled entities; although Tele2 was founded as a subsidiary of the Swedish telecom firm of the same name, it is now divorced from its parent.[113] In turn, these are complemented by a handful of regional service providers that have an outsize market share in particular parts of Russia. These regional service providers include MOTIV in the Urals, Vainakh Telecom in Chechnya, and Tattelecom in Tartarstan.[114]

The Russian mobile service sector exhibits several striking characteristics that distinguish it from the landscape of the overall present-day Russian economy. Unlike many other industries, such as fossil fuel extraction and weapons production, which have been consolidated under the direct or indirect domination of state-owned entities, most of the Russian mobile service industry remains composed of privately held companies.[115] The major exception is Tele2, which is a wholly owned subsidiary of the state-owned Rostelecom, but Tele2 is the smallest of the big four and as a result does not dominate the industry.[116] In the absence of consolidation, the large mobile carrier operators are free to compete vigorously with each other. These firms do not own all of their own infrastructure. Although each owns and operates most of the towers and base stations they use, there are several Russian tower companies that make their business renting towers to the mobile service operators, such as Russkie Bashni, Vertikal', and Servis-telekom.[117] In turn, these towers connect to ground-based cables that are predominantly owned by Rostelecom. Although there are domestic Russian manufacturers of telecommunications

---

[112] Alla Y. Naglis and Xenia A. Melkova, "Telecoms in Russia," King & Spalding LLP, April 29, 2019.

[113] The subscriber base of MTS, Beeline, MegaFon, and Tele2 are overwhelmingly located in Russia and friendly neighboring countries, such as Belarus, Kazakhstan, and Armenia; however, in the past, they have made attempts to enter markets further afield. For instance, Beeline attempted to gain a toehold in the southeast Asian market. See Van Oanh, "Gmobile Replaces Beeline, Targets Mekong," *Saigon Times*, September 19, 2012.

[114] HBR Radiofrequency Technologies, "Russia," webpage, undated-b.

[115] Foreign ownership of telecommunications firms in the Russian Federation is limited by Russian law. Naglis and Melkova, 2019.

[116] HBR Radiofrequency Technologies, "Tele2 Russia," webpage, undated-c.

[117] "Independent 'Tower' Operators Are Headed for the Regions" ["Nezavisimye 'bashennye' operatory poidut v regiony"], IKS Media, December 6, 2018.

equipment, such as Sitronics, these manufacturers provide relatively little of the hardware supporting mobile service to customers.[118] Instead, they are dependent on foreign-designed and foreign-made equipment purchased from suppliers, such as Ericsson and Huawei. The sanctions enacted in response to the 2022 invasion of Ukraine jeopardized the supply of this equipment. Before the invasion, the Chinese firms Huawei and ZTE controlled about 40 to 60 percent of the Russian market for wireless network equipment, while Nokia and Ericsson made up most of the remainder. But Nokia and Ericsson pulled out in response to the invasion, while Huawei and ZTE might also withhold sales of wireless network equipment to Russian customers because of fear of secondary sanctions.[119] Without imported Chinese equipment, it might prove infeasible for Russia to introduce 5G at all.

These considerations are significant for the prospective adoption of 5G in Russia because the mobile provider industry in that country is simultaneously substantially independent of state control and large enough to be a significant commercial force, while lacking the necessary political pull to overcome better-connected entities. Prior to the February 2022 invasion of Ukraine and resulting sanctions, this enabled obstacles to the rollout of 5G in Russia because the so-called power ministries (the Ministry of Defence and the security services) are disinclined to make concessions, such as relinquishing rights to spectrum and allowing the use of standard foreign-made equipment. In practice, this might have made large-scale introduction of 5G technologies in the Russian Federation effectively impossible, even without the crippling effect of international sanctions.

Obstacles to 5G Deployment in Russia

Domestic Obstacles to 5G

In August 2019, the Russian newspaper *Vedomosti* reported that Vladimir Putin had approved a resolution of the Russian Security Council refusing to relinquish the 3.4 to 3.8 GHz band for use by 5G.[120] Officially headed by the Russian president, the Security Council sets Moscow's defense policy and represents the interests of both the Russian military and the security services, such as the Federal Security Service (FSB) and Foreign Intelligence Service (SVR). Although the Russian military had already foreshadowed its disinclination to give up the spectrum used by the emerging global 5G standard, this development must have come as an unpleasant surprise for the Russian mobile carrier industry.

---

[118] Sitronics, "Telecommunications" ["Telekomunikatsii"], webpage, undated.

[119] Ryan McMorrow, Anna Gross, Polina Ivanova, and Kathrin Hille, "Huawei Faces Dilemma over Russia Links That Risk Further US Sanctions," *Ars Technica*, April 1, 2022.

[120] Svetlana Iastrebova, "Putin Doesn't Hand Over Popular 5G Frequencies to Mobile Service Providers" ["Putin ne otdaet operatoram populyarnye chastoty za 5G"], *Vedomosti*, August 15, 2019.

Russian service providers had received permission previously to experiment with 5G service using the 3.4 to 3.8 GHz band, to the extent that MegaFon had set up a preliminary 5G network using it in 11 Russian cities that hosted the 2018 World Cup.[121] This seemed to suggest that the Russian government might adjudicate in the mobile industry's favor when deciding whether to reallocate the spectrum. These firms had highly placed allies within the Russian government. In April 2019, Vice Premier Maksim Akimov made a formal request to Putin to consider relinquishing the 3.4 to 3.8 GHz band for 5G.

Unfortunately, opposition from the highly influential military and space lobbies proved more formidable. That spring, the Ministry of Defence protested that handing over the spectrum would be "premature." In its August 2019 letter, the Security Council insisted not only that Russia's defense could not do without its existing reliance on the 3.4 to 3.8 GHz band but that sharing this spectrum to serve both military and civilian ends was "impossible."[122] As a consequence, Russia's mobile service providers were catapulted into a sea of uncertainty about how to prepare for the next generation of mobile networks.

The Russian communications ministry (Minkomsviazi) proposed an alternative solution: a 5G service using the 4.4 to 4.9 GHz band instead of the 3.4 to 3.8 GHz band. Russian mobile service providers begrudgingly began exploring this option but in general refrained from enthusiastically embracing it. The many deficiencies of the 4.4 to 4.9 GHz band relative to the 3.4 to 3.8 GHz band were a major reason for this hesitancy. Although some other countries were contemplating the use of the 4.4 to 4.9 GHz band as a supplemental spectrum for 5G, no other government planned on using it as the backbone of a 5G standard. The more limited effective range of the higher-frequency band would require a larger number of base stations with a result of higher costs and smaller service areas. Another major obstacle arose in that none of the other countries considering the 4.4 to 4.9 GHz band for 5G also employed the European 4G LTE standard that Russia had adopted. As a consequence, if Russia adopted a 4.4 to 4.9 GHz 5G standard, it would require handsets and other equipment adapted to a combination of a band used nowhere else in the world—with the inevitable result of less choice, smaller economies of scale, and much higher prices.[123]

Given these barriers to building a nationwide 5G network without the 3.4 to 3.8 GHz band, it is hardly surprising that Russian mobile service providers continue to hope that their government will change course and let them build a 5G network in alignment with the emerging world standard. While MTS rolled out a pilot 5G service using the 4.7 to 4.9 GHz band at a handful of well-visited locations in Moscow in March 2021, the Russian telecommunications sector and its

---

[121] Valery Kodachigov, "'Megafon' and 'Rostelekom' Get Frequencies for Fifth-Generation Communications" ["'Megafon' i 'Rostelekom' nashli chastoty dlia piatogo pokoleniia sviazi"], *Vedomosti*, September 20, 2018.

[122] Iastrebova, 2019.

[123] George, Nichiforov-Chuang, and Kolta, 2020, pp. 27–29.

allies in the government continue to lobby for access to the coveted 3.4 to 3.8 GHz band.[124] Predictions that the Kremlin would change course emerged immediately after reports leaked that Putin had agreed to the Russian Defense Council request not to reallocate the 3.4 to 3.8 GHz band, but so far these have yet to translate into any tangible shift in policy.[125]

Russian mobile providers have also been experimenting with millimeter-wave 5G service since the mid-2010s. MTS received a license to begin introducing service on the 24.25 to 24.65 GHz band in July 2020.[126] But later that same year, Russian media reported that the State Radio Frequencies Committee had promulgated a regulation dictating that mobile operators would have to use only domestic-made equipment listed in the Unified Registry of Russian Radioelectronic Production to provide service in this band. This measure was apparently the result of anxieties from the security services about the use of foreign equipment in *critical infrastructure*. But in practical terms, the measure created a serious barrier to the development of millimeter-wave 5G networks because the necessary Russian-made telecommunications equipment was not yet available and might never become available at economic prices.[127]

This is not the sole example of the Russian security services insisting on a potentially crippling "made in Russia" regulation. The FSB has proposed draft legal language that 5G terminal equipment in Russia employ "cryptographic information protection means, including those with confirmation of compliance with the security requirements for KS3 class information established by the federal executive body in the field of security."[128] Depending on how this language is interpreted, it could require the use of nonstandard encryption algorithms that no vendor of 5G equipment is prepared to support because 5G security is supposed to be harmonized in accordance with the TS 33.501 specification (security architecture and procedures for 5G system, release 15).[129]

International Obstacles to 5G

The most-imposing regulatory barriers to Russia's 5G plans, however, might be those brought by its international obligations. Russia is a party to the International Telecommunication

---

[124] "MTS Rolls Out Pilot Consumer 5G Network in Moscow" ["MTS zapustit v Moskve pol'zovatel'skuiu pilotnuiu set' 5G"], Interfax, March 5, 2021; and Andrei Zhriblis, "Is a Federal Antimonopoly Service of Russia Decision Nearing About 5G?" ["Priblizit li reshenie FAS Rossiiu k 5G"], BFM.RU, May 4, 2021.

[125] Zhriblis, 2021; and "Source: Minkomsviaz and Mobile Service Providers Oriented Toward Frequencies from 3.4–3.8 GHz for 5G Development" ["Istochnik: Minkomsviaz' i operatory orientiruiutsia na chastoty 3.4–3.8 GGts dlia razvitiia 5G"], *TASS*, August 15, 2019.

[126] "Government Approves the Use of 24 GHz Radio Frequency Band for 5G Networks" ["Pravitel'stvo utverdilo ispol'zovanie radiochastot 24 GGts dlia seti 5G"], Interfax, May 7, 2021.

[127] Ekaterina Kiniakina, "Russian Mobile Service Providers May Miss Out on the Possibility of Building 5G Networks" ["Rossiiskie operatory mozhet lishit'sia vozmozhnostistroit' seti 5G"], *Vedomosti*, September 20, 2020.

[128] George, Nichiforov-Chuang, and Kolta, 2020, p. 36.

[129] George, Nichiforov-Chuang, and Kolta, 2020.

Union (ITU) and its usage of the EMS is supposed to be harmonized with that of its neighbors. But just as the Russian military uses the 3.4 to 3.8 GHz band for military purposes, so too do the Baltic republics and Finland use the 4.8 to 4.99 GHz band for theirs. These countries employ those frequencies for military aircraft transponders, so under ITU regulations "the use of 4.8– 4.99 GHz frequencies for 5G networks must be negotiated up to 300 km from their air borders and up to 450 km in the case of coastal areas."[130] This is a huge obstacle to the widespread introduction of 5G service based on the higher band because Saint Petersburg falls within this zone, and other countries, such as Ukraine and Georgia, have raised similar objections. In October 2020, the head of Rostelecom warned ominously that these protests could prevent the introduction of 5G service to much of European Russia.[131] Understandably, Russian mobile service providers and other business interests seem to be hoping that this issue will impel a change in government policy to allow the use of the 3.4 to 3.8 GHz band for 5G.[132] So far, the Kremlin has refused to budge, however. In August 2020, Russian officials commented that the uses of the coveted band by the military and security services "are indeed very difficult to convert, and in some cases it is impossible."[133]

Russia's military spectrum usage is also creating a similar problem for some of its neighbors, albeit on a smaller scale. Russia's usage of the 3.4 to 3.8 GHz band for military radars and satellite uplinks in Kaliningrad gives it a veto over the use of that spectrum in about one-third of Lithuanian territory, and this delayed the Lithuanian government from auctioning the 3.5 GHz band for 5G.[134] In February 2021, the Lithuanian government announced that it had given up on direct talks with the Russians on this subject and that it was pushing the issue "at the highest possible level in the ITU community" and the European Postal and Telecommunications Union instead. Russian intransigence was apparently focused on the 3.4 to 3.8 GHz band because the Russians did agree to a Lithuanian request to adjust their usage of the 700 MHz band (which had been used in Russia for digital television service) so that it could be used for 5G.[135]

Implications of Russian 5G Obstacles for U.S. and NATO Military Operations

If these obstacles to 5G utilization along NATO's eastern flank persist, they could pose serious barriers to exploiting the potential of 5G technologies to support military operations.

---

[130] Valery Kodachigov, "Russian 5G Interferes with NATO Airplanes" ["Rossiiskii 5G meshaet samoletom NATO"], *Vedomosti*, February 7, 2021.

[131] "The European Part of Russia May Be Left Without 5G Communications" ["Evropaiskaia chast' Rossii mozhet ostat'sia bez sviazi 5G"], *Russia Today PRIME* [*Rossiia Segodnia PRAIM*], October 14, 2020.

[132] George, Nichiforov-Chuang, and Kolta, 2020, pp. 4–5.

[133] "The European Part of Russia May Be Left Without 5G Communications," 2020.

[134] Maya Guzdar and Tomas Jermalavičius, "Between the Chinese Dragon and the American Eagle: 5G Development in the Baltic States," brief, Tallinn: International Centre for Defence and Security, August 2020.

[135] "Lithuania's 5G Development Hampered by Russian Military Infrastructure," 2021.

Firstly, in circumstances short of war (which might include an intense crisis or *gray zone* conflict), NATO powers will probably still be bound by international agreements governing spectrum utilization, even if the Russian military is defying them. Secondly, if such countries as Lithuania determine that they cannot deploy 5G structure on parts of their territory bordering Russia, that infrastructure will not exist when NATO forces might need to take advantage of it. (At the same time, Russian use of Ukrainian telecom networks during the invasion of Ukraine also indicates that Russian invaders might try to exploit 5G networks in neighboring countries if they can.) However, if Russian policy changes and it harmonizes its 5G policy with the emerging international standard, then these issues might all be rendered moot.

Prospective Russian Military Uses of 5G

Although the Russian military has played an outsize role in discussions of 5G in Russia for the reasons outlined earlier, relatively little has been published in that country about the ways in which the Russian military might itself take advantage of 5G. A search of Russian-language commentary and military science literature identified one major exception—an article that appeared in the fall 2020 issue of the journal *Teoriia I tekhnika radiosviazi* (*Theory and Practice of Radio Communications*). Entitled "The Potential of LTE/5G Mobile Technologies in Military Communications," it was written by M. S. Kondakov and V. V. Solopov, two associates of the "Sozvezdie" enterprise in Voronezh that has supported military EW since Soviet times.[136] The basic thesis of the authors is that commercial off-the-shelf (COTS) 5G technologies will offer immense benefits to Russian military operations in peacetime and outside contested EM environments so long as it can benefit from the economies of scale provided by using equipment based on international standards. Although the article does not say so explicitly, the obvious subtext is that the Russian military should relinquish the 3.4 to 3.8 GHz band for use by 5G.

A Russian View of Military Use of 5G for Noncombat Operations

Although they do not dismiss the battlefield potential of 5G technologies altogether, Kondakov and Solopov make their case almost entirely on the basis of the potential benefits of commercial 5G technology for noncombat applications. As they concede upfront, "In combat conditions . . . very complex and expensive specialized communication systems are needed, which are protected from deliberate interference and electronic warfare equipment."[137] As a consequence of this, the "direct use of civilian mobile technologies, even of the most modern

---

[136] M. S. Kondakov and V. V. Solopov, "The Potential of LTE/5G Mobile Technologies in Military Communications" ["Vozmozhnosti mobil'nykh tekhnologii LTE/5G v voennoi sviazi"], *Theory and Practice of Radio Communications* [*Teoriia i tekhnika radiosviazi*], No. 4, 2020.

[137] Kondakov and Solopov, p. 37.

kinds, is impossible without appropriate enhancements."[138] But in their view, noncombat 5G applications could still be worth it because

> at the same time, in the daily life of the armed forces in peacetime, instead of using complex and expensive specialized military communications equipment, modern commercial, cost-effective mobile communication technologies can be used practically without modifications.[139]

Furthermore, they argue that just because COTS 5G might not be robust enough for a battlefield does not mean it would not be useful in wartime:

> [T]he high frequency efficiency, functionality and manufacturability of civilian standards of mobile communications can be used in the work of troops carried out at sites that are guaranteed to be outside the zones of influence of enemy electronic warfare means, such as remote air bases, ports, forces and means of logistic support, and with appropriate modifications in combat as well.[140]

Kondakov and Solopov suggest that 5G technologies could be modified to work advantageously in a battlespace because their design enables the resiliency associated with the internet protocol. This provides "wide encryption capabilities" and "a high degree of resiliency," in particular "the ability to redistribute available resources and [do so] dynamically."[141] These capabilities make it possible

> to automatically route traffic, make a quick hot swap of all main network elements and . . . balance the load between network components in case of partial failures, suppression by means of electronic warfare or even physical destruction of network infrastructure elements.[142]

For these reasons, they proclaim that "in the opinion of many military experts, with some refinement of the basic LTE/5G technology, it can be effective at the tactical control level in a combat environment."[143]

The authors' case for using commercial 5G systems is both economic and technological: "The use of COTS solutions based on key international standards," they emphasize, "can significantly reduce the cost and time of development and implementation of the latest military communication systems."[144] It can also reduce their "cost of operation, repair, modernization and disposal" while increasing their service life, flexibility and scalability. They attribute these advantages to the

---

[138] Kondakov and Solopov, 2020, p. 37.

[139] Kondakov and Solopov, 2020, p. 37.

[140] Kondakov and Solopov, 2020, p. 37.

[141] Kondakov and Solopov, 2020, p. 45.

[142] Kondakov and Solopov, 2020, p. 37.

[143] Kondakov and Solopov, 2020, p. 37.

[144] Kondakov and Solopov, 2020, p. 37.

methodology of open joint development . . . inaccessible to closed enterprises of the military-industrial complex, with the involvement of the most competent teams of engineers and scientists from leading world companies and laboratories.[145]

As a result, "the capabilities of [LTE/5G standards] are ahead of the capabilities of military systems."[146]

The bulk of Kondakov and Solopov's article is devoted to outlining possible noncombat use cases for 5G technology. They identify three specific "services" they believe COTS 5G networks will provide that they use to frame their analysis. These are enhanced mobile broadband, defined as "high-speed data transmission (up to 1 Gbit/s and above)," massive machine-type communications, defined as "wireless low-speed sensor networks with a density of up to 1 million sensors per sq. km.," and ultrareliable low-latency communication.[147] They structure their overview to describe what these services could do for the Russian military during day-to-day peacetime operations, during combat options (but in rear areas), and during training and education activities. Most of these proposed applications are either mundane, such as faster, more reliable email, or abstracted from 5G technology per se, such as the possible use of augmented reality to facilitate maintenance and training activities.[148]

Others invoke possible intersections with other emerging technologies. For example, the authors suggest, "With the advent of hypersonic weapons . . . the capabilities of ultra-reliable communication with low latency (URLLC) are becoming especially relevant," as these "will make it possible to almost instantly and reliably notify people and automatic control systems about an attack with such a weapon or other sudden danger."[149] This is not wrong, but it is not clear why such an automated warning system needs the specific advantages of 5G, such as low latency as opposed to simple reliability. One plausible 5G application that the authors emphasize at several points is that greater bandwidth availability would facilitate the use of unmanned aerial vehicles (UAVs) and ground-based drones, so long as those were located in a permissive environment.

The authors' conclusions strongly suggest that they are making a veiled argument in favor of reallocating the 3.4 to 3.8 GHz spectrum for 5G. They emphasize that the most important thing is that "multifunction LTE/5G networks provide a cost-effective solution to mobile communication challenges," but that "the economic efficiency of LTE/5G technologies is enabled by the use of commercially available and competitively priced COTS solutions in the open market," whether these are "ready-made or with limited modifications to the needs of individual military

---

[145] Kondakov and Solopov, 2020, p. 47.

[146] Kondakov and Solopov, 2020, p. 47.

[147] Kondakov and Solopov, 2020, p. 39.

[148] Kondakov and Solopov, 2020, pp. 40–41, 46.

[149] Kondakov and Solopov, 2020, p. 43.

customers."[150] These economic benefits hinge on "the use of COTS solutions based on popular international standards."[151] Although the authors do not discuss the fact that Russian 5G policy, as of this writing, rejects those standards, the subtext would be obvious to a Russian reader familiar with the debate about 5G in that country.

Kondakov and Solopov appear to be making the case that the Russian military will gain more than it will lose from handing over the 3.4 to 3.8 GHz band to mobile service providers as

> saving budgetary funds on providing a number of aspects of military communications while simultaneously increasing its functionality will make it possible to reorient the saved funds to achieve other military or civilian goals of the state.[152]

In addition, "the more [military communications] that are transferred to the use of unified standard technologies, the greater the overall synergistic effect will be."[153]

Conclusion

As of this writing, the trajectory of 5G in Russia is fraught with uncertainty because of the hesitation of the Russian military to give up its rights to the contested 3.4 to 3.8 GHz band. Russian mobile service providers point out that attempting to construct a homegrown 5G network using an alternative spectrum might result in Russia missing out on the benefits of the 5G revolution altogether. If this comes to pass, then the Russian military will not be able to exploit the advantages of 5G for its benefit.

Although no evidence has come to light that inconveniencing NATO countries is a goal of Russian policy regarding the 3.4 to 3.8 GHz band, it has already created challenges for such countries as Lithuania that are trying to cultivate 5G networks. In turn, this constrains what NATO can do with 5G in areas close to the Russian border during competition, both because NATO governments feel the need to abide by international agreements governing spectrum usage in peacetime and because NATO militaries cannot take advantage of 5G infrastructure that remains unbuilt because of regulatory conflicts with the Russians. In conflict, the use of private 5G networks is possible, albeit with interference in the mid-band frequencies, while other 5G frequencies will be unaffected.

In any case, should 5G technologies play some role in a conflict with Russia, Russia will attempt to counter 5G technologies in accordance with its distinctive approach to REB. If these measures succeed, the EMS might prove to be a nonpermissive environment for 5G technologies during a conflict with Russia.

---

[150] Kondakov and Solopov, 2020, p. 47.

[151] Kondakov and Solopov, 2020, p. 47.

[152] Kondakov and Solopov, 2020, p. 47.

[153] Kondakov and Solopov, 2020, p. 47.

Chapter 4. Threats to 5G in Europe

Past and Theorized Attacks on Cellular Networks in Europe

Given that 5G as a standard is still being developed, has a variety of different implementations, and as of this writing is only being deployed in a fairly narrow and experimental capacity, it can be difficult to identify vulnerabilities specific to 5G. Even vulnerabilities that are likely to be carried over from existing network infrastructures can be difficult to identify given the lack of real-world penetration testing and regularly updated standards. Considering these factors, we have identified six potential vulnerabilities in 5G. Three of these are potential carryover vulnerabilities from existing network infrastructures that have been previously used by Russia in Europe, and three are potential new vulnerabilities specific to 5G that have been identified in academic cybersecurity literature. Because of the reasons mentioned earlier, it is possible that some or all of these potential vulnerabilities are partially or completely addressed by changes to the 5G standard, the various implementations of 5G infrastructure, or both.

In 2014, Russia engaged in a man-in-the-middle (MitM) attack,[154] which targeted Ukrainian cellular services.[155] The effect of the attack was to intercept and reroute text messages and phone calls to record these conversations.[156] The particular vulnerability used the Common Channel Signaling System Number 7 (SS7) standard and a counterfeit mobile switching center node to gain access to various data normally reserved for security, which could then be used to access other data exchanged between the target and the network node.[157] This vulnerability continues to exist because of current generation (4G and LTE) cellular networks' integration with older network standards and infrastructure (2G and 3G).[158] Because of 5G's requirement to be backward compatible with existing network infrastructure, it is likely that this particular vulnerability will be carried over into 5G networks that are integrated with already vulnerable existing network infrastructure.[159]

---

[154] An *MitM attack* is defined as "An attack in which an attacker is positioned between two communicating parties in order to intercept and/or alter data traveling between them." Paul A. Grassi, Michael E. Garcia, and James L. Fenton, *NIST Special Publication 800-63-3: Digital Identity Guidelines*, Gaithersburg, Md.: National Institute of Standards and Technology, June 2017.

[155] Pavel Polityuk and Jim Finkle, "Ukraine Says Communications Hit, MPs Phones Blocked," Reuters, March 4, 2014.

[156] Polityuk and Finkle, 2014.

[157] Kim Zetter, "The Critical Hole at the Heart of Our Cell Phone Networks," *Wired*, April 28, 2016.

[158] Charlie Osborne, "4G, 5G Networks Could be Vulnerable to Exploit Due to 'Mishmash' of Old Technologies," ZDNet, October 1, 2020.

[159] Osborne, 2020.

Similar to the SS7 switching attack mentioned earlier, traditional cellular jamming attacks will continue to be feasible against 5G networks. Although various changes to the 5G standard, infrastructure implementation, and eventual deployment might make it less vulnerable to jamming attacks, it has already been shown that at least some of these configurations are at risk to jamming attacks. Russia has used such methods on existing cellular network infrastructure various times over the past two decades, but in particular the 2017 attacks on Latvian emergency networks stand out as a particularly important incident.[160] For approximately 72 hours, all nonwired emergency communications networks in western Latvia experienced various degrees of interruption.[161] Although the particular cause of the outage is unknown, the effects are consistent with traditional jamming efforts used by Russia.

In a related case, Russia jammed GPS signals during the 2019 Joint Exercise Clockwork in Northern Norway.[162] This affected both civilian and military ground and air vehicles intermittently for the duration of the exercise. Similar to the attack in Latvia, the exact cause of the attack is unknown, but the effects are consistent with a traditional Russian jamming attack.[163]

In addition to the previous attacks mentioned, we describe three MitM attacks reported by Altaf Shaik and colleagues in 2019.[164] As of this writing, these attacks are only theoretical because they exploit vulnerabilities that are specific to the 5G ecosystem. Fortunately, this makes it possible that future updates to the 5G standard or the particular implementations of 5G will resolve these vulnerabilities.

The first theoretical attack type is *mobile network mapping*, which exploits the lack of required authentication for attribute requests to identify various aspects of a device within the range of a counterfeit base station. With several counterfeit base stations within a given area, it is possible to identify attributes of and approximately locate wireless 5G-enabled devices. Identifiable attributes include the make and model of the device, the operating system it is running, its name, and several other details. With the information derived from this attack, it is possible to identify shifts in the network population of the area around the counterfeit base station and—with enough base stations—also track the movement of these nonlocal devices.[165]

[160] Joseph Trevithick, "Russia Jammed Phones and GPS in Northern Europe During Massive Military Drills," The Warzone, October 16, 2017.

[161] Thomas Grove, Julian E. Barnes, and Drew Hinshaw, "Russia Targets NATO Soldier Smartphones, Western Officials Say," *Wall Street Journal*, October 4, 2017.

[162] Gerard O'Dwyer, "Finland, Norway Press Russia on Suspected GPS Jamming During NATO Drill," Defense News, November 16, 2018.

[163] Alexander Smith, "Norway Calling Out Russia's Jamming Shows European Policy Shift," NBC News, November 24, 2018.

[164] Altaf Shaik, Ravishankar Borgaonkar, Shinjo Park, and Jean-Pierre Seifert, "New Vulnerabilities in 4G and 5G Cellular Access Network Protocols: Exposing Device Capabilities," *Proceedings of the 12th Conference on Security and Privacy in Wireless and Mobile Networks*, May 2019.

[165] Shaik et al., 2019.

Similar to the mobile network mapping attack, the *bidding down* attack also relies on the fact that attribute requests do not require authentication. However, unlike the mobile network mapping attack, the effect is to negate the upload and download speed advantages of 5G. Although bidding down attacks can occur with 4G networks, the specific method that has been theorized by Shaik and colleagues is specific to the IoT-enabling aspects of 5G.[166]

The third theorized attack is a *power drain* attack. This attack involves repeatedly cycling battery-run devices through power saving and regular operation modes to drain their batteries. Like the other theorized attacks, this is also a MitM attack; however, unlike the other attacks, it depends on the 5G-enabled devices' reliance on receiving messages that do not require authentication to determine when they should and should not cycle between different power modes.[167]

Between the different attacks we covered, a few trends emerge. First, all of the theorized attacks and one of the carryover attacks were MitM attacks. This suggests that preventing counterfeit base stations or the direct hacking of legitimate stations could be a strong defense of 5G networks. Second, the theorized attacks all revolve around sending messages that do not require authentication. Finally, because of either integration with existing infrastructure or the physical realities of wireless networks, 5G networks are likely to inherit many of the vulnerabilities of existing cellular networks.

Russian Cellular Electronic Warfare Systems

In addition to past and possible attacks to cellular networks in Europe, Russia has developed several EW systems capable of attacking and exploiting cellular networks in many different ways. Such capabilities could greatly imperil 5G use in a wartime Baltics scenario. Here we describe some of these systems and how they can threaten cellular communications.

Svet-KU (RB-636AM2)

The Svet-KU RB-636AM2 is a signals intelligence, radio, cell phone, and radar jammer with a frequency range between 30 MHz to 18 GHz.[168] It is able to track a variety of emissions and calculate source coordinates and can automatically detect control channels and determine the parameters of radio networks, such as the Global System for Mobile communication (GSM),

---

[166] Shaik et al., 2019.

[167] Shaik et al., 2019.

[168] McDermott, 2017.

Covde Divisio Multiple Access 2000 (CDMA2000), and Universal Mobile Telecommunications System (UMTS).[169] It can also register detected subscriber terminals and block them.[170]

The Svet-KU can be ported on an unmarked Ford Transit van, allowing it to operate covertly on public roads, and a military version has been reported to have been seen in Donetsk.[171] The system can jam and take the place of base stations and be used as virtual cellular stations. The Svet-KU can also geolocate cell phones for targeting. Targeting can be kinetic, or it can influence opposing forces by lowering morale through manipulative text messages. Moreover, it can be used to deliver small disposable ground jammers and jam and spoof GPS.

RB-341V Leer-3

The RB-341V Leer-3 system consists of one command and control (C2) vehicle and three Orlan-10 unmanned aerial systems with jammers.[172] It has an operational range of 120 km, jamming range of 6 km, and frequency range of 800 to 2,000 MHz. It can geolocate and jam cellphones and towers including GSM, 3G, and 4G. The RB-341V can release disposable ground jammers and can appear as a legitimate base station. The system has reportedly been used in Syria to send fake text messages and videos to rebels.[173]

REX-1

The REX-1 is an antidrone ground-based gun built by the ZALA Aero Group, whose parent company is Concern Kalashnikov.[174] It has the ability to block satellite navigation signals, such as U.S. GPS at 2 km, and control the signal between the drone and its operator at 0.5 km.[175] It can suppress signals for various cellular and wireless communications equipment in 800 MHz, 1.8 GHz, 2.1 GHz, 2.4 GHz, and 5.8 GHz.[176]

R-330ZH Zhitel

The R-330ZH Zhitel is a 100 to 2,000 GHz ground-based system capable of geolocating and jamming satellite phones and cell phones. The system consists of a truck, which acts as a control

---

[169] Maksymilian Dura, "Electronic Warfare: Russian Response to the NATO's Advantage? [ANALYSIS]," Defence24, May 5, 2017.

[170] Dura, 2017.

[171] Dura, 2017.

[172] McDermott, 2017.

[173] Keir Giles, "Assessing Russia's Reorganized and Rearmed Military," white paper, Carnegie Endowment for International Peace, May 3, 2017.

[174] ZALA Aero Group, "ZALA Aero," webpage, undated.

[175] Timothy Thomas, *Russia's Electronic Warfare Force: Blending Concepts with Capabilities*, McLean, Va.: MITRE Center for Technology and National Security, September 10, 2020.

[176] "Zala Aero Group REX-1 Drone Jammer," Defense Update, August 26, 2018.

center for operators, and a shelter with four telescopic active phased array transmitter antenna masts.[177] It is reported to have the ability to jam cellular communications to a radius of 30 km.[178]

Leyer-3

The Leyer-3 system blocks cellular communications operating in the GSM-900 and GSM-1,800 bands, shutting down the bands of all cellular networks within a radius of 6 km with the use of a specialized UAV.[179]

Other Possible Cellular Jammers

Other systems have been reported to have the capability to jam and exploit cellular communications, including the Lava-RP, Dzyudoist, Palatin, and the brand new Solyaris-N system. Much of their characteristics and specifications are the product of speculation, but there are probably many more newer systems that have not been reported or observed in the field.

Conclusion

Russians have consistently attacked and exploited cellular networks both in conflict and during peacetime. They have developed multiple EW systems of different sizes and performance, both ground-based and airborne, that have the ability to hold most existing cellular systems at risk. These systems employ multiple ways of denying or exploiting communications, including jamming; employing protocol attacks and MitM attacks; using fake base stations; and geolocating cellular users for targeting and exploiting these networks for information operations by sending seemingly legitimate messages to cellular users for the purposes of enabling psychological operations, spreading misinformation, and creating confusion.

Although it is difficult to predict what vulnerabilities the 5G ecosystem will have and what kinds of attacks it might be susceptible to, Russia has developed multiple systems and used multiple methods to keep all existing cellular systems at risk. It would be safe to assume that this trend will continue throughout 5G and NextG deployment to keep these evolving cellular networks at risk, and Russian systems that put cellular communications at risk will be designed to enable the Russian concepts of holistic and coordinated REB and information warfare.[180]

---

[177] "R-330Zh Zhitel Russian Cellular Jamming and Direction Finding System," webpage, undated.

[178] Thomas, 2020.

[179] Thomas, 2020.

[180] *NextG* refers to future generations of cellular communications beyond 5G.

Chapter 5. Operational Benefits Enabled by 5G

As discussed in the previous chapter, the risks associated with 5G are numerous; however, there are many valuable operational benefits that might be realized in a Baltics scenario. In this chapter, we discuss the opportunities offered by 5G that DoD might want to prioritize in preparation for a future wartime Baltics scenario, as gleaned from a subject-matter expert (SME) workshop. The scenario was divided into discrete tasks that included how 5G supported the execution of that particular task. The scenario was based on a smart logistics vignette that was divided into three parts (for convenience, we grouped *staging* and *onward movement* together):

1. *Reception* is the process of bringing equipment and supplies into ports and airfields in the rear support area in Europe, offloading them, and getting them ready for transit into theater. This relied heavily on 5G for implementing continuous tracking of equipment and supplies using 5G-enabled smart tags. It involved significant use of autonomous vehicles bringing supplies and equipment into ports and airfields and automating the unloading and the movement to different transport modes. This relied on 5G for enabling communications between autonomous vehicles for coordinating these activities and remote communications for coordinating with remote receiving sites. It also involved high-band communications using narrow beam line-of-sight links between equipment to either support jam-resistant communications, very high data rate communications for movement of large data sets between equipment, or both.
2. *Staging and onward movement* is the process of moving equipment and supplies across Europe and supporting and tracking the progress of the movement. This involved the same 5G functionality as in reception but also used 5G signals as an alternative positioning, navigation, and timing (APNT) source in the case of GPS jamming.
3. *Integration* is the process of getting the equipment and supplies for use in theater, including setting them up, maintaining them, creating supply points, and getting supplies to the tactical edge. This includes all previous types of 5G functionality mentioned earlier.

Each operational task was placed in the context of the vignette for different platforms, systems, and echelons. We classified each task in the scenario in one of ten operational contexts generalized and separated from their platform or system and echelon:

- controlling unmanned platforms
- loading and unloading unmanned systems
- transporting systems and supplies
- tracking equipment and supplies
- repairing systems
- arming and refueling
- navigating
- C2
- remote monitoring and control

- transferring data between systems.

We classified the use of 5G for each task in one of five categories of functionality:

- use of smart tags
- high-band communications
- APNT
- communications between unmanned vehicles
- remote communications.

Three two-hour workshops were held with participants who had expertise in at least one of the following areas:

- military operations (sustainment and logistics)
- technical expertise (communications, including 5G)
- autonomous vehicles, including unmanned aircraft systems
- artificial intelligence
- EW and cyber.

Participants were presented with details of the smart logistics scenario, described in Appendix A, and were then asked to qualitatively score on a scale of low, medium, or high each step of the scenario with their expert judgement on the following: operational impact of 5G, resilience with 5G, and 5G uniqueness, as defined in Table 5.1.[181]

Table 5.1. 5G Scoring Metrics

Feature	Low = 1	Medium = 2	High = 3
Operational utility of 5G	5G has minimal impact on any meaningful operational metric, such as kill ratio, speed of maneuver, reduction in manpower, etc.	5G has measurable impact on one or more operational metrics	5G enables operational capabilities not previously possible with standard systems
Resilience to red attack	Blue can easily be attacked over large distances remotely with fairly unsophisticated systems	Blue can be attacked in standoff distances (e.g., line of sight) with typical jamming, EW, and cyber weapons with reasonable risks and resources	Blue can only be attacked within close proximity of blue forces and would require significant resources, planning and/or risk
5G uniqueness	Task can be done with current systems by changing configurations and tactics, techniques, and procedures (TTP)	Task can be done with upgrades to other communication technologies, but 5G might offer an advantage in cost or lower complexity	Task can only be achieved with 5G; it would require significant investment to achieve with other technologies

---

[181] During the workshop, we asked participants to score tasks on non-5G operational impact. This was done to ensure that scores were not conflating operational impacts from 5G with operational impacts from other innovative technologies, such as autonomous vehicles or artificial intelligence. Because those scores are not relevant to the research, they are not presented here.

We presented each task in the context of the three phases of the vignette and the capabilities that 5G would offer in support of each task. After presenting each task, we asked participants to enter their scores into a spreadsheet. The numerical score assigned for each was 1 for low, 2 for medium, and 3 for high, for the purposes of deriving a numerical average score across respondents. We analyzed the data and classified the scores by the functionality 5G would enable. We then calculated the sum of all three scores as the total score. This should not be interpreted as an in-depth analysis of the impact of 5G but as trends in the minds of different SMEs from various backgrounds. Results should be viewed as general potential trends that should be investigated further and used to prioritize the focus of future efforts and more in-depth analysis. Tasks and functionalities that received low scores should not be discarded because these could still be valuable and valid uses of 5G, and the lower scores could reflect either the way that tasks and functionalities were presented in the workshop or the context of the specific operational mission. The lower scores should not be interpreted as relevant in the context of different missions, operational tasks, and platforms.

In the next three sections, we present the average scores and trends associated with them. We also discuss potential underlying facts and aspects of the operational context that might explain these trends. The explanations are not intended to accurately reflect the thought process of the SMEs but are considerations that should be viewed as a starting point for future research.

Reception Phase

Figure 5.1 shows the results for the reception phase of the vignette. We plotted the average operational utility scores for each task (one dot represents one discrete task) versus the total scores across all three metrics (left panel of Figure 5.1). We also plotted the total score for each task versus the functionality that 5G was enabling in support of each task (right panel of Figure 5.1). The left panel shows two groupings: one with operational utility and total scores above 2 and 6 respectively (in the ellipse) and one group with low scores. The right panel shows the total scores against the 5G functionality and highlights in the ellipses the ones that correspond to the high-scoring tasks. We see that smart tags in this phase of the vignette scores consistently high. High-band communications scored consistently low. Unmanned communications scores varied; two tasks scored high and one scored low. The large number of unmanned loading and unloading systems and unmanned transportation systems that need to coordinate in this phase might be reflected in these higher scores. The one that scored low was related to communications between noncrewed surface vessels, and it could be interpreted as 5G not being available across enough of the ocean waterways in the vignette. The one remote communication task scored low, potentially reflecting the variety of communications options available in the rear support area and in a relative sanctuary from Russian interference. Smart tags are clearly the 5G functionality that was considered to be more important across all three metrics overall, followed by certain types of communications between unmanned vehicles.

Figure 5.1. Average Scores for the Reception Phase

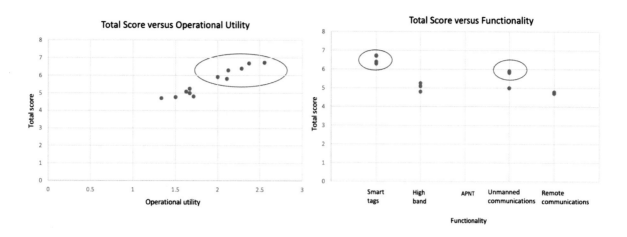

Staging and Onward Movement

Figure 5.2 shows the results for the staging and the onward movement phase of the vignette. We plotted scores as described in the preceding section. The left panel shows two groupings, just as before, but not as separated as in the previous section. The right panel shows the total scores against the 5G functionality and highlights in the ellipses the ones that correspond to the high-scoring tasks. The trends for this phase of the vignette look different. Smart tags still score high, but not consistently. This might be because there are generally fewer tags collocated in restricted areas while the equipment and supplies are in transit across Europe. Remote communications also score high; looking at the particulars of these tasks, we note that they represent remote communications with allies and local support contractors. High-band, unmanned communications and remote communications scored low, as did APNT. There was one high-band task that was related to the downloads of large data sets, which will be further discussed in the next section.

Figure 5.2. Average Scores for the Staging and Onward Movement Phase

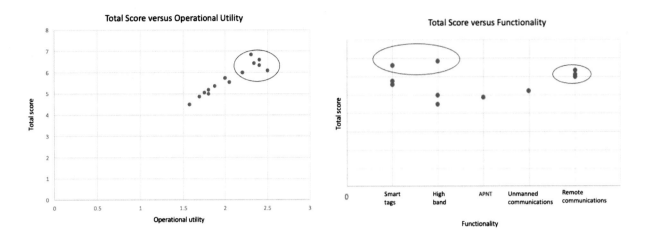

Integration

In Figure 5.3, we present the results for integration phase of the vignette. We plotted scores as described in the preceding section. The left panel shows two less distinct groupings. The right panel shows the total scores against the 5G functionality and highlights in the ellipses the ones that correspond to the high-scoring tasks. We note not only some of the same trends highlighted in the staging and onward movement phase but also a few differences. Smart tags again consistently scored higher just as in the reception phase, perhaps reflecting the higher concentration of equipment and supplies in the forward integration areas in theater. There were no remote communications tasks in this phase. There is one high-band communications task and one unmanned communications task that also scored high. These two tasks were related to the download of large data sets, just as in the one case in the previous section. It would appear that downloading large data sets might also score high, regardless of the types of communication used. APNT again scored low as in the previous phase.

Figure 5.3. Average Scores for the Integration Phase

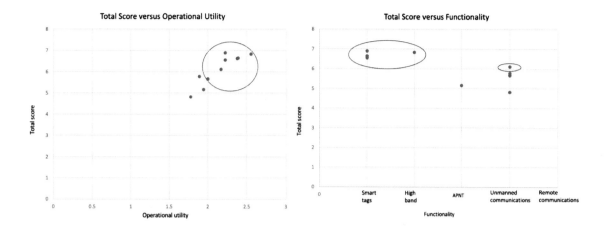

All Scores Combined

Figure 5.4 combines all scores across all phases plotted as operational utility versus resilience and operational utility versus uniqueness. We see that operational utility and resiliency scores seemed to be correlated. When resiliency scored high, effectiveness generally did so as well (Figure 5.4, left panel). Uniqueness scores were consistently lower, scoring 2 or above in very few cases (Figure 5.4, right panel, denoted in the ellipse).

Figure 5.4. Scores for All Phases

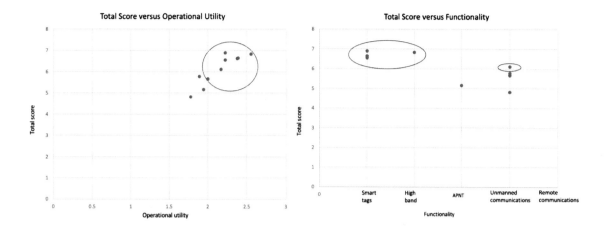

Conclusions

Overall, the results show that the phase and operational context matter. Different phases have different geographical and physical constraints, different threat environments, and different operational priorities. When trying to assess the impact of using a 5G capability, one must carefully consider the operational context and the physical and environmental constraints, among other concerns. We also noticed that operational utility and resiliency scores seemed to be correlated. Uniqueness scores were consistently lower, scoring 2 or above in very few cases.

Smart tags score high consistently across all phases of the vignette but generally score lower during staging and onward movement. This is possibly due to the lower density of equipment and supplies during transit. Other functionalities will score differently in various phases of the vignette, so the exact operational context matters. During reception and integration, communications between unmanned vehicles during loading and unloading operations seemed to score higher. During staging and onward movement, communications with allies and contractors at remote locations seemed to score higher.

There was one outlier in staging and onward movement regarding the download of large data sets with high-band 5G. Two more outliers were observed in the integration phase regarding the transfer of sensor data between operators and across networked vehicles using high-band 5G and between unmanned vehicles. These three outliers seem to be related and scored higher, and they might deserve some closer investigation. A summary of score ranges can be seen in Table 5.2 (there were cases for which there was only a single task for some functionalities and phases, and hence a single value is reported).

Table 5.2. 5G Scoring Ranges by Phase and Functionality

	Reception	Onward Movement	Integration
Smart tags	6.30–6.72	5.75–6.60	6.55–6.89
High-band communications	4.80–5.25	4.50–6.85	6.83
Unmanned communications	5.00–5.91	5.20	4.82–6.11
Remote communications	4.70–4.78	6.00–6.35	N/A
APNT	N/A	4.87	5.17

NOTE: N/A = not applicable.

As was mentioned earlier, these should not be viewed as an in-depth analysis of the use of 5G within the operational context but merely as starting points for future research. Tasks within which 5G did not score well should not be discarded. Other factors, such as the mission context within which the SMEs scored the use of 5G or the way the task and use of 5G was explained in the workshop, could have had a large impact on the scoring.

Chapter 6. Recommendations and Conclusions

5G promises faster communication speeds, lower latency, and higher capacity than have yet been possible with previous networks. It can enable advancements, such as mobile edge computing, software-defined radios and networks, and inexpensive directional antennas. These can support the downloading of massive data sets, the coordination of armed unmanned aircraft system swarms, and the tracking of millions of small IoT devices. With the emergence of network slicing in 5G, there might be increased security as well. Such innovations are likely to greatly alter the tempo and TTP of military operations in the European theater and elsewhere. Next, we present recommendations to the Office of the Under Secretary of Defense for Research and Engineering (OUSD R&E) as it considers future conflicts in Europe in the 2030 time frame.

Recommendations

Future OUSD R&E 5G research for supporting military logistics should be focused on smart tags, high-band 5G, remote communications, and unmanned communications. These were the main 5G opportunities we found in our idealized Baltics smart logistics scenario drawing on expert consensus. In particular, smart tags might be useful in all phases of wartime logistics by tracking thousands of items, such as materiel and emergency supplies, in near real-time. This information can then be sent back to artificial intelligence algorithms for monitoring and maintenance. High-band 5G will be useful for transferring sensor data between operators and unmanned vehicles and networks. Remote communications will be useful in establishing robust C2 architecture and communicating across allies and support contractors. Communication for the coordination of unmanned vehicles during loading and unloading also seems to be a promising feature of 5G.

However, we noticed that the SMEs scored these opportunities differently depending on the phase of the scenario. Smart tags scored well during all phases (RSOI). High-band communications scored well during the staging, onward movement, and integration phases. Remote communications with allies scored well during the staging and onward movement phases. Communications for unmanned systems scored well during the reception and integration phases. Additional research should consider the operational context of each of these features.

OUSD R&E should also expand 5G research into different operational scenarios and vignettes, such as urban operations, humanitarian assistance and disaster recovery, and special forces operations. Although we recommend focusing future research in the areas noted earlier in the context of Baltics wartime logistics, others also deserve attention and might perform well in different contexts. For example, APNT did not score well in our smart logistics scenario. This is likely because its benefits will be reduced if Russia can use fake 5G emitters. We noticed an interesting outcome in which unmanned vehicles had high scores, but unmanned ships did not. This is likely because there is less infrastructure for 5G over water, and because high-band frequencies have shorter range. However, these features might score differently in other operational scenarios.

Develop mitigation approaches for countering Russian exploitation and denial of 5G. It is likely that during a conflict, Russia will attempt to exploit our use of 5G to gain operational advantage against blue forces. To mitigate this, the United States and its allies should be prepared to defend against jamming, EW threats, protocol attacks, fake base stations, and information warfare messaging. Any vulnerabilities in both existing networks and previous-generation networks should be mitigated so that they are not carried over into the 5G ecosystem.

Track frequency developments across allies and their borders with Russia. Frequency disputes have delayed 5G deployment in countries that neighbor Russia. It is still unclear when or whether these disputes will be resolved. The ITU, a part of the United Nations, would presumably negotiate this conflict. The issue is expected to be presented at a commissioner conference in 2022 and at the World Radio Confederation in 2023.[182] Assuming these 5G frequency disputes remain unresolved in 2030, the DoD and allies should consider alternative methods to conduct C2, communication, and coordination in those countries, particularly for operations that might rely heavily on high bandwidth and low latencies in border areas.

Assess potential interoperability issues across providers and countries. A large-scale war in Europe will require active coordination and communication between Joint partners, allies, and host nations. Blue forces are likely to face interoperability issues, particularly when moving across borders. As 5G is rolled out and new equipment and concepts of operation are developed, mitigating these challenges will be paramount. Training and equipping the force with interoperability in mind will be resource intensive, but will be necessary to ensure successful military operations and realize the full array of 5G benefits.

Work with European allies to create a common roadmap of military uses of 5G. The 5G ecosystem is not yet fully deployed in Europe, but is expected to be rolled out within the next decade in the ten countries examined in this research. There are several ongoing initiatives of trials that test 5G in urban corridors and cross-country motorways, railways, waterways, and ports. As the development of the 5G ecosystem continues, a roadmap detailing the military uses across allies and host nations will be necessary to seamlessly carry out operations.

Conclusion

Existing modes of communication and military logistics in the European theater can be slow and cumbersome. It is difficult to predict future capabilities and use cases, but 5G is likely to transform wartime logistics by enhancing the way the U.S. military communicates with partners,

---

[182] Daly, 2021.

tracks and transports equipment and personnel, and transfers sensor data. Although the progress of the 5G rollout within each country differs drastically, and a wide variety of threats—including jamming, employing protocol attacks and MitM attacks, using fake base stations, and geolocating cellular users for targeting and exploiting—are likely to jeopardize the 5G ecosystem, we identified several opportunities that will allow the DoD and U.S. allies to prioritize further research. We believe that these recommendations will improve the development of 5G for military use and pave the way for future operational success.

Appendix A. Detailed Description of Smart Logistics Vignette

The smart logistics operations vignette was developed to investigate aspects of how 5G technology can enhance logistics force projection capabilities to deploy a designated joint force task force in support of a U.S. European Command Baltics scenario. This is intended to separate key elements (e.g., vessel discharge and speed to closure) that might affect the results and findings from elements that would not have an impact. The detailed vignette is described here to illustrate the variety of factors considered in improving logistics processes, particularly RSOI, to project the joint force in support of large-scale combat operations in the Baltics region. Figure A.1 depicts an overview of the RSOI process.

This appendix contains the logistics vignette broken down into three parts:

- situation and mission
- RSOI concept and overview
- logistics performance improvement (existing versus 5G-enhanced practices).

Figure A.1. Reception, Staging, Onward Movement, and Integration Process Overview

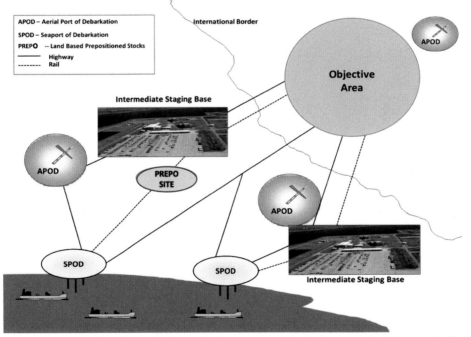

SOURCE: Army Techniques Publication 3-35, *Army Deployment and Redeployment*, Washington, D.C.: Headquarters, U.S. Department of the Army, March 2015, Figure 4-1.

From its inception as U.S. Army Europe (USAREUR) in 1952 and through the end of the Cold War, one of the primary responsibilities for USAREUR was to plan for the execution of combat in Western Europe. A large part of this mission involved receiving forces arriving from the continental United States (CONUS) and integrating them into the combat formations prepared to defeat a potential Soviet invasion—a process known as RSOI.[183] Today's Army organizations aligned to fight inside the multi-domain concept face immense challenges, which evolve constantly depending on mission, enemy, terrain and weather, troops and support available, time available, and civil considerations and the permissiveness of the environment. These variables play a decisive role in determinations regarding the RSOI process. The purpose of RSOI is "to build the combat power necessary to support the Ground Component Commander's (GCC) concept of operation."[184] The complexities of moving large elements from the home station via air and seaports of embarkation to the air and seaports of debarkation (SPODs) present a variety of opportunities for friction and delay. This RSOI process in Europe presents challenges specific to the theater as they relate to movement of a large joint force from reception locations throughout Western Europe to final integration in eastern Poland.

Situation and Mission

This vignette describes existing and future concepts of how projection of forces involving personnel and equipment occurs against a near-peer competitor that has invaded a U.S. partner nation. Notionally, U.S. European Command deploys U.S. forces into the Baltics region using organic and commercial power projection assets aligned in comparison with future unmanned and autonomous transportation capabilities enabled by 5G. In line with a classic fait accompli scenario, the adversary conducted an exercise close to the border, following claims of threats to its borders and the partner nation. The exercise forces then conducted a limited-scope invasion. The invaded country was unable to repel the larger and more capable force. The U.S. response will require setting the theater using forces already present to blunt the attack, massing contact layer forces in the region, and conducting RSOI to surge additional forces from CONUS, which can marry up with U.S. European Command and U.S. Marine Corps prepositioned equipment in theater.

The joint deployment and redeployment processes, illustrated in Figure A.2, consist of four phases: planning, predeployment activities, movement, and joint RSOI. Both processes are similar; however, each has distinct characteristics. These phases are iterative and might occur simultaneously throughout an operation.[185] *Planning* is the first phase of the process and occurs

---

[183] The EU term for this type of operation is *reception, staging and onward movement* or *reception, staging, onward movement, and integration*. See European External Action Service, *EU Concept for Reception, Staging, Onward Movement and Integration (RSOI) for EU-Led Military Operations*, Brussels: Council of the European Union, EEAS 0078/12, May 11, 2012.

[184] Army Techniques Publication 3-35, 2015, p. 4-1.

[185] Joint Publication 3-35, *Deployment and Redeployment Operations*, Washington, D.C.: U.S. Joint Chiefs of Staff, January 2018.

throughout the entire process during both the preparation and execution functions of the exercise. *Predeployment activities* are actions taken before movement to prepare for a deployment operation. *Movement* involves the activities to physically move joint forces from origin to destination and consists of movement from origin to port of embarkation; activities at the port of embarkation; movement from port of embarkation to port of debarkation, including movement and transit through intermediate locations as required; and movement from port of debarkation to final destination. *Joint RSOI*, the final phase, is an essential process that transitions deploying or redeploying forces. *Deployment* consists of personnel, equipment, and materiel arriving into a theater of operations task organized to meet the combatant commander's operational requirements. *Redeployment* completes the movement of forces to home station or a demobilization site, concluding the end-of-mission or exercise.

Figure A.2. Joint Deployment and Redeployment Processes

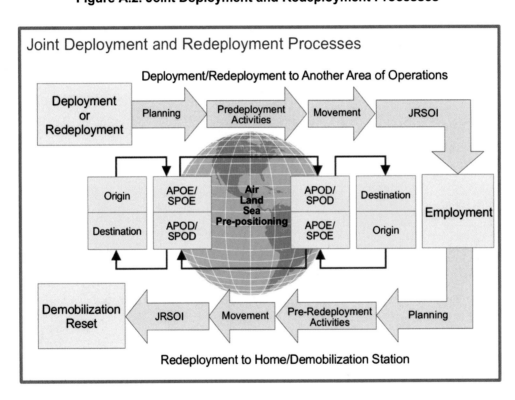

SOURCE: Joint Publication 3-35, 2018, Figure I-3.
NOTE: APOD = airport of debarkation; APOE = airport of embarkation; JRSOI = joint reception, staging, onward movement, and integration; SPOE = seaport of embarkation.

Deployment planning and execution decisions are based on the anticipated operational environment to be encountered in the operational area.[186] Understanding the operational

---

[186] Deployment operations are the activities required to plan, prepare, and move forces and materiel from home station to a destination to employ an operational capability required to execute a mission. The focus of these

environment helps commanders anticipate the results of various friendly, adversarial, and neutral actions and how they affect operational depth and reach and mission accomplishment.

Threat

The environment is nonpermissive. The adversary possesses a robust electronic and cyber warfare capability and the most likely course of action is to conduct cyber EW attacks against U.S. forces as they deploy into theater while preparing defensive positions with conventional forces and long-range strike systems. The adversary might also employ forces to conduct raids against key U.S. and partner nation targets. These actions are meant to dissuade U.S. forces from projecting combat power in the region.

Friendly Forces

Having anticipated the potential for a crisis in the region, the United States established prepositioned forces consisting of the U.S. Army, U.S. Air Force, and U.S. Marine Corps. Additionally, limited forces stationed across the European theater inside U.S. partner nations are included in a scenario involving the following forces: U.S. Army Brigade Combat Teams and supporting units; present rotational CONUS-based forces; U.S. Marine Corps forces consisting of a Marine Expeditionary Unit; and U.S. Air Force elements stationed in Germany ordered to deploy three deployable airfield-in-a-box systems.

Surge forces from CONUS to support the operation consist of one armored brigade combat team, one Stryker Brigade Combat Team, one infantry brigade combat team, one Army Combat Aviation Brigade, one Fires Brigade (Field Artillery—High Mobility Artillery Rocket System), and one logistics Sustainment Brigade. These forces, along with assigned forces in Europe, will deploy to the Baltic region via intrastrategic lift using commercial assets across rail and road networks and strategic sealift from U.S. ports along with joint U.S. Marine Corps prepositioned forces entering the northern Baltic region.

Concept of Operations

In response to orders to deploy, the scenario concept of operation will consist of four phases focusing on force projection:

- **Phase 1: Set the theater.** Deploy forces to complicate enemy targeting, build days of supply for U.S. forces, and relocate prepositioned equipment to ISBs. Establish C2 capabilities to move major combat units to designated areas throughout the theater.
- **Phase 2: Ensure local access.** Move major combat forces to ISBs or TAAs.

---

operations is to globally position forces in time to conduct military activities, including campaigns and major operations, and to respond to other contingencies.

- **Phase 3: Staging and onward movement of forces.** The United States conducts joint force operations to task and organize forces in the Baltic region.
- **Phase 4: Consolidation and integration at TAAs.** U.S. forces consolidate and continue to flow heavier forces into occupied territories.

Reception

Reception is the first stage of the RSOI process: Proper reception of personnel and equipment arriving by sea and air sets the stage for success and the rapid integration of forces later in the process. Air and sea movements have specific challenges to overcome for the deploying and theater support units. Reception encompasses all necessary functions to properly receive and account for all personnel and equipment arriving at the air and SPOD.[187] Reception sets the tone for the entire RSOI operation occurring at air and seaports and railhead or truck depots. An example timeline for reception of a Stryker Brigade Combat Team via airlift might require approximately one to three weeks.[188] Reception operations occur at seaports, depots, and airfields. Doctrinally, airfield operations are joint resourced and begin with the Air Force Tanker Airlift Control Element in charge of ramp and airfield operations, while the Army operates the Arrival and Departure Airfield Control Group, which is tasked with receiving and processing the incoming personnel and equipment. Port operations consist of joint, multinational, and commercial operations.[189] Considerations specific to the SPOD operations include infrastructure analysis, determinations of the number of berths and piers, offload cranes, and draft depth. These play integral roles in the type of ship that a particular port can handle. Port operations involve tasks shared across multiple agencies, such as the partner nation and U.S. service contracts executed by U.S. Transportation Command Surface Deployment and Distribution Command, power and energy needs, and road and rail network output. Vessel discharge operations are the primary focus of SPOD operations as part of force projection and the RSOI process. Both port and airfield operations involve unloading equipment into the staging areas for convoy configuration, rail uploading, or even cross loading back onto military aircraft.

The two biggest constraints in SPOD operations are space and time. There is a limited amount of space available in staging yards for military equipment, and competition for berths with commercial vessels is an ever-present concern. Even during wartime operations, major ports still need to conduct commerce to keep the economy moving. Balancing port operations regarding military operations and being able to deconflict those operations with civilian traffic as much as possible is a political-military policy issue that can have a huge impact on the capacity of the port to receive and process forces. U.S. doctrine directs that once all equipment is offloaded from vessels and aircraft, they will be staged and organized for travel via rail or

---

[187] Joint Publication 3-35, 2018, p. VI-2.

[188] Joint Publication 3-35, 2018, p. 3-1.

[189] Army Techniques Publication 3-35, 2015, p. 4-6.

ground convoy and/or combined with commercial line haul to an ISB for the next step in the RSOI process.

Electric autonomous ship transport enhanced by 5G capabilities has the potential to support task-organized type loading to eliminate reorganization of equipment and unit systems at the port. Another key element that electric vessels provide is minimal ballast constraints because they can use their battery bank as ballast. This changes the systemic loading by type of equipment by weight rather than by system lethality. Equipment configured on autonomous platforms might benefit from smart pallet or container configuration for vertical takeoff and lift aviation platforms. Through these enhancements and changes, expanded port of entry options might happen closer to the ISB or TAA.

Staging

In today's existing staging phase of the RSOI operation, unit personnel and equipment are physically linked up in the vicinity of the ISB. However, during port and airfield operations, staging equipment awaiting onward movement begins as soon as the first piece of equipment is offloaded.[190] Here, the unit's equipment is to be arranged to facilitate the mode of transport designated to deliver to a forward location. The accuracy in which the commercial industry performs this has much needed room for improvement. It is not until the unit is staged at the ISB that equipment is reassembled and tested, the unit conducts theater-specific training, and the unit is scheduled to move to the TAA for integration.[191]

Onward Movement

Onward movement timelines can be the most challenging and time consuming for planning purposes. Many factors affect the time it takes to close the force (movement) from certain distances from the ISB or ISBs to the TAA. This involves several convoys, units to close on the TAA, traffic management along the lines of communication, blackout times, reconfiguring of equipment, download and upload of equipment at the ISB site, etc. A brigade-sized force on average completes travel from the ISB to the TAA for integration within five to seven days. Of all the stages of RSOI in Europe, onward movement is perhaps the most challenging. What is unknown at this point is the final destinations for integration of forces. Certain planning factors have been ascertained to help in the planning of an onward movement operation. A quick overview of some transportation and infrastructure considerations will now be explored. Once all tasks encompassed in staging are complete, the unit is ready to move to the TAA. As with staging, onward movement can occur more than once, depending on the size of the theater and the distance from the ISB to the TAA. Onward movement is not only a joint but also a combined

---

[190] Army Techniques Publication 3-35, 2015, p. 4-7.

[191] Army Techniques Publication 3-35, 2015, p. 4-8.

undertaking; coordination with multinational forces across theater infrastructures can and must occur. Successful coordination and deconfliction of routes and infrastructure (used for storage, maintenance, and billeting of personnel) not only with joint and combined forces but also with civilian traffic is paramount to manage unit movement requirements against the route capabilities. Units need to be successful in onward movement operations in three key areas: movement control, use and understanding of transportation infrastructure, and security.[192]

Transport Mode (Rail and Road)

In a theater of operations, the Army is tasked with providing movement control of joint land transportation movements.[193] Lines of communication must be carefully maintained and remain uncongested. Creative logistics and transportation solutions need to be considered, such as the following:

- Accelerate operational decisionmaking at the operational and tactical level.
- Transfer sensor data between operators and uninhabited vehicles and network vehicles.
- Use battlespace sensor data to inform integration processes of C2 and security.
- Transition to begin combat operations (configure for combat).
- Conduct drone (heavy) lift capabilities to deliver personnel, fuel, and ammunition.
- Ensure days-of-supply commodity storage levels are met.
- Employ smart-transport, repair, and supply chain programs that reduce by half the footprint of contractors, fuel storage sites, repair parts, and U.S. personnel during defensive and offensive operations.
- During relief in place operations, observe the incoming unit, and offer suggestions based on mission and lessons learned.
- Enable augmented maintenance environments that provide advance additive manufacturing part production and robotic repair.
- Reduce risk in conduct arming, fueling, and fixing equipment during combat by integrating drone and unmanned aircraft system (vertical takeoff and lift) capabilities across the corps and division.
- Enhance force projection at all operational objectives using GPS-guided systems.
- Regenerate the force during combat losses (materiel and personnel).
- Reduce latency across logistics and maintenance to reduce time required to sustain the Multi-Domain Task Force across multiple countries.
- Support autonomous battlefield recovery capabilities for the United States and NATO.

---

[192] Army Techniques Publication 3-35, 2015, p. 4-10.

[193] Field Manual 4-01, *Army Transportation Operations*, Washington, D.C.: U.S. Department of the Army, April 2014, pp. 1–6.

Integration

The overall process of integration involves units that initially arrive at a specified location as well as the process of departing units into and out of the tactical theater; it occurs when all units have closed on the TAA.

Appendix B. 5G Rollout in Countries of Interest

This appendix discusses strategies, future plans, and objectives in Estonia, Latvia, Lithuania, Belgium, Finland, Germany, the Netherlands, Norway, Poland, and Sweden. These countries were chosen as relevant to a Baltic states military scenario that involves both the forward tactical areas and the support rear areas required to move equipment and supplies into theater. In the sections that follow, we provide a brief description of each country's 5G national strategy or equivalent, when available; status of spectrum license auctions; and carriers and network coverage.

Estonia

Estonia's Ministry of Economic Affairs and Communications released its *Digital Agenda 2020 for Estonia: Updated 2018 (Summary)* outlining its national strategy for developing a "mature and secure environment for the widespread use and development of smart ICT solutions."[194] Broad strategic future steps related to 5G were as follows: developing an Estonian 5G activity plan, making radio frequency resources available in the 700 MHz and 3.5 GHz frequency bands, and participating in international 5G cooperation projects with the Nordics, Baltics, and Poland. The Estonian *5G Roadmap Through 2025* establishes Estonia's goals of 5G coverage in all major cities by 2023 and transport corridors by 2025, spectrum auctions for 3.4 to 3.8 GHz licenses in spring 2019, and 694 to 790 MHz licenses in the first half of 2020, and public consultations for market interest in the 24.25 to 27.5 GHz frequency band.[195] Estonia suspended the 3.4 to 3.8 GHz auction scheduled for April 2019 because of a complaint from Levikom Estonia—an IoT and fixed-wireless internet service provider—that three licenses favor Estonia's three major cellular communications companies: Elisa Estonia, Tele2 Estonia, and Telia Estonia.[196] As of January 2022, Estonia still has not auctioned the spectrum; however, the Minister of Entrepreneurship and Information Technology, Andres Sutt, announced in December 2021 that Estonia will auction three licenses for 130 MHz each in the 3,410 to 3,800 MHz spectrum.[197] Estonia still has neither auctioned nor scheduled auctions for the 700 MHz and 26 GHz spectrum bands. A public consultation in 2019 resulted in Estonia learning that there is little

---

[194] Republic of Estonia Ministry of Economic Affairs and Communications, *Digital Agenda 2020 for Estonia: Updated 2018 (Summary)*, Tallinn, 2019.

[195] European Commission, "Broadband in Estonia," webpage, last updated February 4, 2022a.

[196] European 5G Observatory, "Estonia," webpage, last updated 2021a; and Pujol et al., 2021.

[197] James Barton, "Estonia to Put Three 3.5GHz Licences Up for Auction Next Year," Developing Telecoms, December 17, 2021b.

market demand for the 26 GHz band. Because of these delays, Estonia lags behind EU averages for 5G readiness and 5G coverage in the 2021 DESI indicators.[198]

Telia Estonia launched the first 5G commercial services in November 2020 in Estonia's three largest cities—Tallinn, Tartu, and Parnu—using existing spectrum and Ericsson's dynamic spectrum sharing technology.[199] Telia Estonia rapidly expanded its 5G coverage to include 50 locations by June 2021;[200] it also included 77 locations across 14 cities by July 2021.[201] Elisa Estonia and Tele2 Estonia have not launched commercial 5G services, but both are preparing for their 5G rollout and spectrum licenses. Elisa Estonia and Nokia announced in December 2021 that they entered into a five-year agreement for nationwide deployment of Nokia's 5G radio access network equipment, after Estonia completes its 3.5 GHz auction.[202] In January 2022, Tele2 announced that it also entered into an agreement with Nokia as its 5G radio access network equipment vendor in the Baltics.[203] Telia and Tele2 began considering Nokia and Ericsson as their hardware providers after the Estonian parliament approved an amendment to its Electronic Communications Act granting the Estonian government the power to force communication companies to provide the government information on the hardware and software on the communication companies' networks.[204] Although not mentioned by name in the amendment, it is known as the Huawei Law because it came on the heels of a joint declaration between the United States and Estonia on evaluation network security of 5G providers.[205]

Latvia

Latvia has not published a national 5G strategy but began commercial 5G deployment in January 2020. Latvia held its first spectrum auction for 100 MHz of the 3.4 to 3.8 GHz band in November 2017. LMT was the sole participant and received a ten-year license for two 50-MHz blocks—3,400 to 3,450 MHz and 3,650 to 3,700 MHz—in the spectrum valid from January 1, 2019, to December 31, 2028.[206] Latvia held a subsequent auction for a ten-year license for 3,550

---

[198] European Commission, undated-d, p. 8.

[199] "Telia Launches First Public 5G Network in Estonia," Telia Company, November 10, 2020; and Catherine Sbeglia Nin, "Telia Turns on 5G in Estonia with Ericsson," RCR Wireless, November 11, 2020.

[200] "Telia's 5G Network Available at over 50 Locations in Estonia," *Baltic Times*, June 14, 2021.

[201] Hala Turk, "Telia Eesti 5G Network Reaches 77 Regions in Estonia," Inside Telecom, July 26, 2021.

[202] Nokia, "Nokia Wins Five-Year 5G Deal with Elisa Estonia as Sole RAN Vendor," press release, Espoo, Finland, December 20, 2021.

[203] Anne Morris, "Tele2 Picks Nokia for 5G RAN in Baltics," Light Reading, January 4, 2022a.

[204] European Commission Directorate-General for Communications Networks, Content and Technology, *5G Observatory Quarterly Report 13 Up to October 2021*, Brussels, October 2021; and European Parliamentary Research Service, "Example: Estonia," webpage, undated.

[205] European Parliamentary Research Service, undated.

[206] "LMT Secures 5G-Compatible Spectrum," CommsUpdate, December 11, 2017.

to 3,600 MHz in September 2018, which Tele2 Latvia won. This license also lasts from January 1, 2019, to December 31, 2028.[207] Latvia held a 700 MHz spectrum auction in December 2021.[208] LMT, Tele2, and Bite Latvia all won paired blocks during the auction. LMT acquired the 713 to 723 MHz and 768 to 778 MHz channels, Tele2 the 703 to 713 MHz and 758 to 768 MHz channels, and Bite Latvia the 723 to 733 MHz and 778 to 788 MHz channels.[209] The licenses are for 20 years and valid from February 1, 2022, until January 31, 2042.[210] LMT and Tele2 also won 17-year licenses for unpaired spectrum blocks. LMT acquired the 748 to 758 MHz channel and Tele2 acquired the 738 to 748 channel, and these licenses last from February 1, 2025, until January 31, 2042.[211] With these auctions, Latvia has completed auctions for the 700 MHz and 3.4 to 3.8 GHz spectrums.

There are three 5G operators in Latvia: LMT, Tele2 Latvia, and Bite Latvia. LMT launched commercial 5G services in July 2019, making it the first operator to do so—not only in Latvia but also the Baltics.[212] LMT began developing its 5G network after winning auctions in the 3.4 to 3.8 GHz band in 2017. LMT had deployed 100 5G base stations by December 2021.[213] LMT did so in cities including (but not limited to) Riga, Kuldīga, Ventspils, Ogre, Ikšķile, Sigulda, and Skrīveri.[214] Tele2 Latvia launched 5G services starting in January 2020 and claims that after securing licenses during the 700 MHz spectrum auction, it now has everything in place to commence 5G rollout across Latvia.[215] Both LMT and Tele2 used their 3.5 GHz spectrum to launch their respective 5G services.[216] Bite Latvia began operating a single 5G base station in June 2019 and expanded to seven base stations in January 2021 but does not plan wider deployment until 2023.[217]

---

[207] "Tele2 Latvia Snaps Up 3.5GHz Spectrum for 5G," CommsUpdate, September 17, 2018.

[208] "Tele2 and Bite Claim Success in 700MHz Auction Ahead of Final Approval of Results," CommsUpdate, December 20, 2021.

[209] James Barton, "All Available Spectrum Sold in Latvia's 700MHz Auctions," Developing Telecoms, January 4, 2022.

[210] Tele2, "Tele2 Wins Latvian 5G Frequencies," press release, Stockholm, December 16, 2021.

[211] "Latvian Mobile Communications Operators Pay a Pretty Penny for 5G Frequency Rights," Baltic News Network, December 27, 2021.

[212] "LMT Will Install 100 5G Base Stations This Year," LMT, September 8, 2021; and "LMT Flicks the Switch—Becomes First Mobile Operator to Launch 5G Internet in the Baltics," *Baltic Times*, July 26, 2019.

[213] "LMT Deploys 100th 5G Base Station," CommsUpdate, December 10, 2021.

[214] "LMT Will Install 100 5G Base Stations This Year," Labs of Latvia, October 22, 2021.

[215] Tele2, 2021.

[216] European Commission Directorate-General for Communications Networks, Content and Technology, 2021.

[217] James Barton, "Bite Pledges 5G Investment as It Switches on New Sites," Developing Telecoms, January 21, 2021a.

Lithuania

The Lithuanian Ministry of Transport and Communications approved the 5G Development Guidelines of Lithuania for 2020 to 2025 in June 2020.[218] Lithuania's plan had four stages: beginning 5G connection and development in 2021; providing 5G availability in at least one of its largest cities by 2022; achieving uninterrupted 5G function in Lithuania's five largest cities (Vilnius, Kaunas, Klaipeda, Siauliai, and Panevezys) by 2023; and providing 5G connectivity along the international transport corridors Via Baltica and Rail Baltica, roads of national importance, and railways and ports by 2025.[219] Lithuania also set a goal to allocate pioneer bands before 2021. Lithuania has not completed auctions for any pioneer bands yet. In October 2021, Lithuania announced its first auction for any pioneer band, 700 MHz. Lithuania will award three blocks, one 2x10 MHz wide blocks and two 2x5 MHz wide blocks in the 713 to 733 MHz and 768 to 788 MHz bands. Interested parties had to submit their auction documents by January 25, 2022.[220]

Lithuania has three operators planning to offer 5G services: Bite Lithuania, Tele2 Lithuania, and Telia Lithuania. Telia Lithuania launched a trial 5G mobile network in November 2020 with a temporary frequency license from Lithuania's Communications Regulatory Authority;[221] it is in the 3.5 GHz frequency band and an existing 2.1 GHz frequency band.[222] Telia Lithuania stated that it had 11 base stations in Vilnius, Kaunas, and Klaipėda, but the frequencies were for testing noncommercial 5G communication services.[223] In October 2021, Telia Lithuania announced that it had installed 250 base stations since that summer and that it planned to launch 5G commercial services as soon as it could purchase spectrum.[224] Bite Lithuania and Tele2 Lithuania have not launched 5G services yet.

The Lithuanian parliament prohibited "unreliable" manufacturers and suppliers from Lithuania's electronic communications markets in May 2021 and further required technology screening to maintain national security.[225] Similar to Estonia, operators began replacing Huawei equipment with products from Ericsson or Nokia. Telia Lithuania entered into a five-year partnership with Ericsson as its sole 5G network provider in November 2020 and the two companies plan to modernize approximately 3,000 base stations within three years of that

---

[218] "Lithuania Approved the 5G Development Guidelines for 2020–2025," Lithuanian Ministry of Transport and Communications, June 3, 2020.

[219] "Lithuania Approved the 5G Development Guidelines for 2020-2025," 2020.

[220] Chris Dziadul, "First 5G Auction for Lithuania," Broadband TV News, October 26, 2021.

[221] "Telia Lietuva Launches Trial 5G Network," CommsUpdate, November 11, 2020.

[222] Anne Morris, "Telia Gets 5G Ball Rolling in Lithuania with DSS," Light Reading, January 14, 2022b.

[223] "Telia Is Launching 5G Network in Lithuania," Telia Company, November 10, 2020.

[224] "Telia Is Ready for Commercial 5G Launch: Installed 250 Base Stations," Telia Company, October 13, 2021.

[225] "Lithuania Bans 'Unreliable' Technologies from Its 5G Network," LRT English, May 25, 2021.

agreement.[226] Tele2 partnered with Nokia for its 5G radio access network hardware in Lithuania.[227]

Belgium

The Belgian Institute for Postal Services and Telecommunications (BIPT) published its communication concerning the introduction of 5G in Belgium in September 2018.[228] At publication, BIPT planned to allocate user rights for the 700 MHz and 3,400 to 3,800 MHz bands by fall 2019, the 26 GHz band starting in 2021, and the 31.8 to 33.4 GHz and 40.5 to 43.5 GHz bands starting in 2022.[229] In October 2018, BIPT released the details of the 700 MHz auction setting the reserve price for a 5 MHz block, 20-year license at 20 million euros.[230] BIPT planned to auction the 400 MHz between 3,400 and 3,800 MHz based on two lots: 360 MHz of the spectrum for which no entity has user rights and 40 MHz for which two entities hold regional user rights. As of this writing, Gridmax and Citymesh—two Belgian telecommunications companies—hold regional licenses through 2025. For the 26 GHz band, the public showed little interest during BIPT's 2019 public consultation, so BIPT has not set a date for its auction based on lack of market demand.[231] Despite these plans, Belgium has not permanently allocated or auctioned these spectrums as of January 2022, largely because of disagreement between the federal and regional governments over the distribution of auction proceeds,[232] increases in the number of antennas, and the extent of rural area coverage.[233] BIPT granted temporary user rights for the 3,600 to 3,800 MHz band in July 2020 to support the initial 5G rollout. Four entities received user rights. Cegeka received 40 MHz and Orange Belgium, Proximus, and Telenet each received 50 MHz within the spectrum.[234] Belgium is ranked below average for 5G readiness and

[226] Ericsson, "Lithuania Added to Telia and Ericsson Nordics and Baltics 5G Alliance," press release, Stockholm, November 3, 2020.

[227] Tele2, "Tele2 Chooses Nokia for 5G RAN Rollout in the Baltics," press release, Stockholm, January 4, 2022.

[228] The BIPT is "the federal regulatory body responsible for regulating the electronic communications market, the postal market, the electromagnetic spectrum of radio frequencies and the radio and television broadcasting in the Brussels-Capital Region." See Belgian Institute for Postal Services and Telecommunications, "BIPT," webpage, undated-b; and Belgian Institute for Postal Services and Telecommunications, *Communication by the BIPT Council of 10 September 2018 Regarding the Introduction of 5G in Belgium*, Brussels, September 10, 2018.

[229] Belgian Institute for Postal Services and Telecommunications, 2018.

[230] European 5G Observatory, "National 5G Strategies: Belgium," webpage, last updated 2021b.

[231] Belgian Institute for Postal Services and Telecommunications, *Annual Report 2019*, Brussels, May 29, 2020.

[232] European Commission, undated-b.

[233] "Belgian 5G Auction Terms Receive Government Approval," CommsUpdate, November 25, 2021.

[234] Belgian Institute for Postal Services and Telecommunications, 2020.

coverage within the EU largely because of these auction delays.[235] The Belgian government approved the terms of the 5G auction in November 2021, and BIPT will likely hold the auction in the second quarter of 2022.[236]

Proximus launched the first 5G commercial services in Belgium in April 2020.[237] It uses 2.1 GHz because Proximus already had user rights for that spectrum.[238] Proximus states that it is technologically ready, and its network is capable of operating 5G; however, it must wait for the government to organize the spectrum auction, grant licenses, and update the electromagnetic field standards.[239] Proximus decided to use the 2.1 GHz spectrum until those conditions are met. According to the 5G Observatory, Proximus did not deploy 5G in the Brussels-Capital Region because of its higher electromagnetic field standards.[240] The majority of its network covers the Flanders Region, particularly the central areas of Ghent and Antwerp.

Telenet started deploying 5G using its temporary license for the 3.6 to 3.8 GHz band in December 2021,[241] but Telenet stated that its full 5G rollout can only happen after the spectrum auction.[242] It plans to focus its rollout in Leuven, Antwerp, and the Belgian Coast and will cover its entire network with 5G by 2025.[243] Orange Belgium activated its 5G network in Antwerp in February 2022 using its temporary license in the 3.6 to 3.8 GHz band.[244] Cegeka has not launched 5G commercial services as of March 2022.

Finland

Finland became the first EU member state to assign the 700 MHz frequency band in 2017; however, it noted the band can provide 5G services that need wide coverage, but it is not a proper 5G band because it lacks the bandwidth necessary for high-speed 5G data transfer.[245] To promote the implementation of 5G, Finland's Ministry of Transport and Communications

[235] Axon Partners Group Consulting, *Study on the Financial and Environmental Impact of 5G and a Fourth Mobile Network Operator in Belgium*, Brussels: Belgian Institute for Postal Services and Telecommunications, April 2021; and European Commission, undated-b.

[236] "Belgian 5G Auction Terms Receive Government Approval," 2021.

[237] Proximus, "Launching 5G in Belgium," fact sheet, undated-b.

[238] Pujol et al., 2021.

[239] Promixus, "Frequently Asked Questions About 5G," webpage, undated-a.

[240] Pujol et al., 2021.

[241] Juan Pedro Tomás, "Telenet to Start Gradual Deployment of 5G in Belgium Next Week," RCR Wireless, December 3, 2021c.

[242] "Telenet to Launch 5G in Belgium Using Provisional Spectrum," European 5G Observatory, December 8, 2021.

[243] "Telenet to Launch 5G in Belgium Using Provisional Spectrum," 2021.

[244] Geert van der Klugt, "Orange Belgium Activates 5G in Antwerp, Rest of Belgium to Follow," Techzine, February 11, 2022.

[245] Ministry of Transport and Communications, 2019, p. 20.

(Traficom) published its *Digital Infrastructure Strategy 2025* in October 2018, outlining objectives and measures.[246] By the time of the *Digital Infrastructure Strategy 2025*, Finland had already auctioned the 3.5 GHz spectrum (3,410 to 3,800 MHz) for the construction of nationwide wireless broadband and allowed operators to begin building networks starting on January 1, 2019.[247] Finland's plan scheduled the issuing of user rights for the 26 GHz spectrum in spring 2020. Finland met all of the European Commission's goals for allocating user rights of the pioneer bands by June 2020.

The 5G Observatory reported that Finland's 3.5 GHz spectrum auctions concluded in October 2018 and Telia Finland, Elisa Finland, and DNA each won a 130 MHz license within the spectrum. Telia Finland, Elisa Finland, and DNA received the 3,410 to 3,540 MHz, 3,540 to 3,670 MHz, and 3,670 to 3,800 MHz frequency bands.[248] These licenses are valid for 15 years starting January 1, 2019. Traficom announced that these three companies won the auction within the 26 GHz spectrum.[249] Each received an 800 MHz band with Elisa, Telia Finland, and DNA receiving rights to the 25.1 to 25.9 GHz, 25.9 to 26.7 GHz, and 26.7 to 27.5 GHz frequency bands, respectively, and the operating licenses are valid from July 1, 2020, until December 31, 2033.[250]

Elisa Finland, Telia Finland, and DNA each offer commercial 5G services within the 3,500 MHz band.[251] Elisa Finland claimed to be the first operator to launch commercial 5G and sell subscriptions in June 2018.[252] Elisa Finland has now expanded to cover nearly 4 million Finns in January 2022 since its initial launch, which is an increase from approximately 2 million in March 2021.[253] Telia Finland started offering 5G services in November 2019. In November 2021, Telia Finland launched a 5G standalone core network, which "operates as a fully independent network, providing for higher capacity, increased security, lower latency, and network functionality" as opposed to nonstandalone networks that share signaling and transfer priorities with 4G networks.[254] DNA started offering mobile 5G in January 2020 using 3.5 GHz and by November 2020 the network expanded to 76 municipalities and 1.5 million people.[255] DNA activated its 700

---

[246] Ministry of Transport and Communications, 2019.

[247] Ministry of Transport and Communications, 2019.

[248] "3.4–3.8 GHz Auctions in Finland," European 5G Observatory, October 2, 2018.

[249] Traficom Finnish Transport and Communications Agency, "Spectrum Auction for 26 GHz Frequency Band," webpage, June 11, 2020.

[250] Colin Mann, "Finland Concludes 26 GHz 5G Spectrum Auction," Advanced Television, June 9, 2020.

[251] HBR Radiofrequency Technologies, "Finland," webpage, undated-a.

[252] Elisa, "Elisa First in World to Launch Commercial 5G," press release, Helsinki, June 27, 2018.

[253] Elisa, "CEO's Review Q4 2021," webpage, undated.

[254] Telia Company, "Telia Company Launches 5G SA Core Network in Finland—the First Commercially Available 5G SA Network in the Nordic and Baltic Region," press release, Solna, Sweden, November 10, 2021b.

[255] Pujol et al., 2021, p. 25.

MHz frequency band in November 2021 and now its network covers 3.5 million people or 64 percent of the population, an increase from 21 percent in November 2020.[256]

Germany

The Federal Ministry of Transport and Digital Infrastructure published its *5G Strategy for Germany* in September 2017, which outlined five action areas: step up network rollout; make frequencies available based on demand; promote cooperation among telecommunications and user industries; conduct targeted and coordinated research; and initiate 5G for towns and cities.[257] As of this writing, Germany is a leader in 5G within the EU, scoring 100 percent in 5G readiness and 18 percent in 5G coverage compared with the EU average of 51 percent and 14 percent, respectively. Germany assigned the 700 MHz frequencies in June 2015 and the 3.4 to 3.7 GHz in June 2019, reserved 100 MHz of the 3.7 to 3.8 GHz spectrum for regional and local purposes, and is making 26 GHz available based on market demand.[258] The German national regulatory authority, Bundesnetzagentur, allocated the rights of 3.7 to 3.8 GHz to approximately 120 entities for 5G campus and industrial networks.[259]

Germany was the first country in Europe to open the 700 MHz band to auction in January 2015.[260] Three companies bid and received licenses in the 700 MHz frequency from Bundesnetzagentur in June 2015: Deutsche Telekom, Telefónica Germany, and Vodafone Germany.[261] These three companies also bid and received licenses in the 3.5 GHz spectrum along with a new entrant, 1&1 Drillisch.[262] Deutsche Telekom and Vodafone Germany received licenses for 90 MHz, Telefónica Germany received 70 MHz, and 1&1 Drillisch received 50 MHz of the 3.5 GHz spectrum.[263] Vodafone Germany initially launched its 5G network in July 2019 and uses the 3.6 GHz, 1.8 GHz, and 700 MHz bands.[264] Deutsche Telekom reaches

---

[256] "DNA's 5G Network Coverage Now Reaches More than 3.5 Million Finns in Nearly 140 Towns as a Consequence of a New 5G Frequency Band," DNA, December 13, 2021.

[257] Federal Ministry of Transport and Digital Infrastructure (Germany), *5G Strategy for Germany: A Scheme to Promote the Development of Germany to Become a Lead Market for 5G Networks and Applications*, Berlin, 2017.

[258] Pujol et al., 2021, p. 91.

[259] European Commission, undated-f, p. 10.

[260] Jörn Krieger, "Germany Opens 700 MHz Frequency Auction," Broadband TV News, January 29, 2015.

[261] "German Spectrum Auction Raises EUR5.08bn," CommsUpdate, June 22, 2015.

[262] Douglas Busvine, "Germany Raises 6.55 Billion Euros in Epic 5G Spectrum Auction," Reuters, June 12, 2019.

[263] "German 5G Spectrum Auction Concludes," HBR Radiofrequency Technologies, June 12, 2019.

[264] Juan Pedro Tomás, "Vodafone Reaches over 45 Million People in Germany with Its 5G Network," RCR Wireless, January 13, 2022a.

approximately 90 percent of the German population and uses two frequencies: 3.5 GHz and 2.1 GHz.[265] Telefónica launched its 5G service in October 2020 using its 3.6 GHz frequency.[266]

The Netherlands

In 2018, the Netherlands published the *Dutch Digitalisation Strategy 2018–2021: Getting the Netherlands Ready for the Digital Future*,[267] the *Connectivity Action Plan*,[268] and *5G and the Netherlands Authority for Consumers and Markets*.[269] These documents outlined goals and tasks related to 5G rollout. The Netherlands also published *The Dutch Digitalisation Strategy 2021*.[270] The Dutch government's objective is to make 5G widely available by 2025 and make sufficient frequency spectrum available for 5G networks.[271] The Netherlands Ministry of Economic Affairs and Climate Policy (EZK) planned to auction spectrum within the 700 MHz band in late 2019 or early 2020.[272] The EZK did not have plans to auction spectrum within the 3.5 GHz or 26 GHz bands when it issued its strategy in 2018. The Netherlands previously issued licenses for spectrum in the 3.5 GHz band through 2022 and 2026. Additionally, Dutch intelligence and security agencies use the 3.5 GHz band for national security purposes, which required the EZK to assess whether it could make spectrum available.[273]

EZK auctioned spectrum within the 700, 1,400, and 2,100 MHz bands in July 2020.[274] Three telecom providers acquired frequencies in these bands: KPN, T-Mobile Netherlands, and VodafoneZiggo. All three providers won 2x10 MHz in the 700 MHz band; KPN and VodafoneZiggo won 1x15 MHz and T-Mobile won 1x10 MHz in the 1,400 MHz band; and all

---

[265] Juan Pedro Tomás, "Deutsche Telekom's 5G Network Covers over 90% of Germany's Population," RCR Wireless, February 16, 2022b.

[266] Telefónica, "Focus on 'Pure 5G' over 3.6 GHz Frequency Fast O₂ Network Expansion: 50 Per Cent 5G Coverage by the End of 2022," press release, Munich, Germany, October 5, 2021.

[267] Secretary of State for Economic Affairs and Climate Policy, Minister of Justice and Security, and State Secretary of the Interior and Kingdom Relations, *Dutch Digitalisation Strategy 2018–2021: Getting the Netherlands Ready for the Digital Future*, The Hague, The Netherlands, September 30, 2019.

[268] Ministry of Economic Affairs and Climate Policy, *Connectivity Action Plan*, The Hague, The Netherlands, July 2018.

[269] Authority for Consumers & Markets, *5G and the Netherlands Authority for Consumers and Markets*, The Hague, The Netherlands, December 12, 2018.

[270] Secretary of State for Economic Affairs and Climate Policy, Minister of Justice and Security, and State Secretary of the Interior and Kingdom Relations, *The Dutch Digitalisation Strategy 2021*, The Hague, The Netherlands, June 22, 2021.

[271] Secretary of State for Economic Affairs and Climate Policy, Minister of Justice and Security, and State Secretary of the Interior and Kingdom Relations, 2019, p. 36.

[272] Authority for Consumers & Markets, 2018, p. 13; and Secretary of State for Economic Affairs and Climate Policy, Minister of Justice and Security, and State Secretary of the Interior and Kingdom Relations, 2021.

[273] Ministry of Economic Affairs and Climate Policy, 2018, p. 16.

[274] "KPN, T-Mobile and VodafoneZiggo Acquire Frequencies in Dutch Mobile Communications Auction," 2020.

three won 2x20 MHz in the 2,100 MHz band.[275] The 700 MHz band licenses required the telecom providers to cover at least 98 percent of the Netherlands by July 28, 2022.[276] The Netherlands planned to auction spectrum in the 3.5 GHz band in April 2022 but faced a legal challenge from Inmarsat—a British satellite company—in March 2021. Inmarsat uses part of the spectrum for emergency satellite communications, but the Dutch national spectrum plan reserves the spectrum for mobile services starting in September 2022, which would prevent Inmarsat from using it. Inmarsat obtained a court injunction that suspended the Dutch government's auction and the EZK delayed the auction for an unknown period.[277] The Netherlands held a public consultation for the 26 GHz spectrum that closed in February 2020 but has not scheduled a 26 GHz auction yet.

KPN, T-Mobile Netherlands, and VodafoneZiggo launched 5G services in the Netherlands. VodafoneZiggo began offering services in April 2020, covering over half of the country initially and expanding to cover more than 80 percent of the population by July 2020.[278] VodafoneZiggo is using the 800 MHz, 1.8 GHz, 2.1 GHz, and 2.6 GHz frequencies.[279] T-Mobile and KPN both activated their 5G networks in July 2020. By October 2020, T-Mobile became the first carrier in the Netherlands to offer nationwide coverage, which is 90 percent of the population.[280] KPN launched the same day as T-Mobile and covered approximately half of the Dutch population.[281] All three carriers are using the 700 MHz band for their 5G networks. According to the 2021 DESI country profile, the Netherlands had the second highest 5G coverage in the EU at 80 percent of total households and 30 percent of rural households.[282]

Norway

The Norwegian Communications Authority announced the planned spectrum auction schedule through its 5G spectrum roadmap in June 2019. The Norwegian Communications Authority auctioned 20-year spectrum licenses for the 700 MHz band and 13.5-year licenses in the 2.1 GHz band in June 2019.[283] Telenor Norway, Telia Norway, and Ice received 2x10 MHz

---

[275] "Dutch Multi-Band 5G Auction Raises EUR1.23bn," CommsUpdate, July 22, 2020.

[276] European Commission, undated-i, p. 8.

[277] "Inmarsat Wins Injunction Against Dutch Refarming of 3.5 GHz Band," 2021.

[278] "Vodafone First to Offer Nationwide 5G," Vodafone, July 20, 2020.

[279] Anne Morris, "VodafoneZiggo Uses Dynamic Spectrum Sharing to Get Ahead with 5G," Light Reading, April 28, 2020.

[280] "T-Mobile First Dutch Telecom to Offer Nationwide 5G Coverage," NL Times, October 27, 2020.

[281] Juan Pedro Tomás, "T-Mobile Netherlands, KPN to Launch Commercial 5G Networks Next Week," RCR Wireless, July 24, 2020b.

[282] European Commission, undated-i, p. 7.

[283] "Nkom Announces Results of 700MHz, 2100MHz Spectrum Auction in Norway," European 5G Observatory, June 14, 2019.

blocks in the 700 MHz band and Ice also received 2x15 MHz in the 2.1 GHz band.[284] The 700 MHz spectrum licenses stipulate that the holders must provide mobile broadband access to at least 40 percent of the population within five years and holders had the opportunity to receive a discount if they agreed to cover prescribed areas. Telenor and Telia agreed to obligations of covering "European roads and the coastal road from Mo I Rana to Bodo'" and "selected railway lines," respectively.[285] Norway completed its 2.6 to 3.6 GHz band auction in September 2021, assigning 190 MHz in the 2.6 GHz band and 400 MHz in the 3.6 GHz band. Telia won 2x30 MHz, Telenor 2x40 MHz, and Altibox 50 MHz in the 2.6 GHz band.[286] In the 3.6 GHz band, Ice won 80 MHz, Altibox 100 MHz, Telenor 1,120 MHz, and Telia 100 MHz.[287] Altibox, Telia, and Telenor received discounts in their auction prices because each committed to make broadband services with greater than 100 Mbps download available in rural areas.

Poland

Poland's Ministry of Digital Affairs released the *5G Strategy for Poland* in January 2018; as of this writing, Poland planned to launch 5G services through the 700 MHz band in at least one major city by the end of 2020, assign spectrum licenses in the 3.4 to 3.8 GHz and 26 GHz bands by the end of 2021, and cover major transport paths by 2025.[288] Poland's goals included accelerating and improving 5G investment, reducing infrastructure maintenance costs, making spectrum available in a timely manner, ensuring security, and reviewing telecommunications regulatory frameworks.[289] The Polish Office of Electronic Communications opened its first frequency spectrum auction for the 3.6 GHz band in March 2020 but had to cancel it in May 2020 because of the COVID-19 pandemic. In November 2020, the Council of Ministers updated its National Broadband Plan, delaying distribution of the 3.6 GHz bands until August 2021 and announcing the auctioning of the 700 MHz band by July 2022.[290] As of this writing, the 26 GHz auction is scheduled for December 2022.[291] Poland has yet to assign any 5G spectrum as of January 2022. According to DESI indicators, Poland's 5G readiness was at 0 percent and its 5G

---

[284] "Nkom Announces Results of 700MHz, 2100MHz Spectrum Auction in Norway," 2019.

[285] "Nkom Announces Results of 700MHz, 2100MHz Spectrum Auction," CommsUpdate, June 6, 2019.

[286] Norwegian Communications Authority, "The Norwegian 5G Auction Has Concluded," webpage, September 30, 2021.

[287] Norwegian Communications Authority, 2021.

[288] European 5G Observatory, "National 5G Strategies: Poland," webpage, last updated 2021c.

[289] Dominik Kopera, "'5G for Poland' Strategy," briefing slides, Warsaw: Ministry of Digital Affairs, undated, p. 7.

[290] Damian Karwala, "5G Regulation and Law in Poland," CMS, undated.

[291] Pujol et al., 2021, p. 48.

coverage was at 10 percent in April 2021, both of which fall below the EU averages of 51 percent and 14 percent, respectively.[292]

Plus became the first operator to offer 5G services in Poland in May 2020. Plus used its 2.6 GHz spectrum and 100 base stations to offer 5G to approximately 900,000 people in seven cities: Gdansk, Katowice, Lodz, Poznan, Szczecin, Warsaw, and Wroclaw.[293] By January 2021, Plus increased its coverage to more than 7 million people and infrastructure to 1,000 base stations.[294] Plus planned to continue its 5G rollout to 11 million people, 1,700 base stations, and 150 cities by the end of 2021.[295] Play launched its 5G services in June 2020 with 50 base stations in 16 cities.[296] Orange Poland activated its 5G network in July 2020 using its 2.1 GHz frequency.[297] The Orange Poland network covers approximately six million people and nine of Poland's 14 provinces. T-Mobile Poland became the fourth operator to offer 5G services in July 2021 and has rapidly expanded its network. T-Mobile stated that it launched 120 to 140 new base stations each week between July and September 2021, had a total of 2,840 5G base stations across Poland, and covered approximately 25 percent of the country.[298] T-Mobile also uses the 2.1 GHz band for its 5G services.

Sweden

In March 2017, Sweden's Ministry of Enterprise and Innovation released *A Completely Connected Sweden by 2025—a Broadband Strategy*, which outlined a vision of "access to high-speed broadband in all of Sweden and access to reliable and high-quality mobile services."[299] To date, Sweden held its 700 MHz auction and 3.6 GHz auctions and is considering a 26 GHz auction. The Swedish Post and Telecom Authority finished the 700 MHz auction in December 2018. Telia and Net4Mobility—an infrastructure joint venture between Swedish telecommunications companies Tele2 and Telenor—won a 2x10 MHz block license and two 2x5 MHz blocks, respectively.[300] The Swedish Post and Telecom Authority concluded the 3.5 GHz spectrum auction in January 2021. Three companies won licenses: Hi3g Access, Net4Mobility,

---

[292] European Commission, undated-k, p. 7.

[293] Pujol et al., 2021.

[294] "Plus Extends 5G Coverage. Over 7 Million of Poles Within the 5G Network Coverage," Grupa Polsat Plus, January 5, 2021.

[295] "Plus Extends 5G Coverage. Over 7 Million of Poles Within the 5G Network Coverage," 2021.

[296] Pujol et al., 2021, p. 31.

[297] "Orange Launches 5G Network," ICT Market Experts, June 5, 2020.

[298] "T-Mobile 5G Covers 25% of Poland," CommsUpdate, September 16, 2021.

[299] Government Offices of Sweden, undated, p. 6.

[300] Swedish Post and Telecom Authority, "Assignment in the 700 MHz Band," webpage, December 11, 2018.

and Telia.[301] Hi3G won 100 MHz in the 3,400 to 3,500 MHz range, Net4Mobility won 100 MHz in the 3,620 to 3,720 MHz range, and Telia won 120 MHz in the 3,500 MHz to 3,620 MHz range.[302] Sweden is slightly below the EU average in 5G readiness at 49 percent (compared with 51 percent) and about equal to the EU average for 5G coverage in DESI indicators.[303]

As of this writing, Tele2, Telia Sweden, Tre Sweden, and Telenor Sweden offer commercial 5G services in Sweden. Tele2 launched its 5G services in May 2020 using 80 MHz that Tele2 had purchased in the C-band after delays in the 3.5 GHz spectrum auction.[304] Like Tele2, Telia Sweden started offering 5G services in May 2020 with 15 base stations in Stockholm.[305] By July 2021, Telia had expanded its network to 22 cities. Telia expects to cover more than 90 percent of the Swedish population by 2023 and 90 percent of Sweden's geographical area and 99 percent of the population by 2025.[306] Tre Sweden began offering 5G in June 2020, primarily in Stockholm but also in Malmö, Helsingborg, Lund, Västerås, and Uppsala.[307] Lastly, Telenor launched its services in October 2020 starting with central Stockholm.[308] The Telenor network uses 80 MHz in the 3.7 GHz band that it shares with Tele2 from their joint venture.[309] Telenor plans to cover 99 percent of Sweden's population by 2023.

---

[301] Swedish Post and Telecom Authority, "Assignment in the 3.5 GHz and 2.3 GHz Bands," webpage, last updated January 21, 2021.

[302] Swedish Post and Telecom Authority, 2021.

[303] European Commission, undated-l, p. 7.

[304] Tele2, "Tele2 5G Is Here," webpage, undated.

[305] Telia Company, "Telia Sweden's First Major 5G Network Up and Running in Stockholm," press release, Solna, Sweden, May 24, 2020.

[306] Telia Company, "The Need for Speed: Telia and Ericsson Beat Swedish 5G Speed Record," press release, Solna, Sweden, July 5, 2021a.

[307] Juan Pedro Tomás, "Three Sweden Launches 5G in Six Cities," RCR Wireless, June 16, 2020a.

[308] Juan Pedro Tomás, "Telenor Launches 5G in Stockholm, Targets Nationwide Coverage by 2023," RCR Wireless, October 30, 2020c.

[309] Tomás, 2020c.

Abbreviations

3GPP	3rd Generation Partnership Project
4G LTE	fourth generation long-term evolution
5G	fifth generation
5G PPP	Fifth Generation Infrastructure Public Private Partnership
APNT	alternative positioning, navigation, and timing
BIPT	Belgian Institute for Postal Services and Telecommunications
C2	command and control
CAM	connected and automated mobility
CCD	camouflage, concealment, and deception
CONUS	continental United States
COTS	commercial off-the-shelf
COVID-19	coronavirus disease 2019
DESI	Digital Economy and Society Index
DoD	U.S. Department of Defense
EM	electromagnetic
EMS	electromagnetic spectrum
EU	European Union
EW	electronic warfare
EZK	The Netherlands Ministry of Economic Affairs and Climate Policy
FSB	Federal Security Service
GPS	Global Positioning System
GSM	Global System for Mobile communication
HET	heavy equipment transport
IBCT	infantry brigade combat team
ICT	information and communication technology
IoT	internet of things
ISB	intermediate staging base
ISR	intelligence, surveillance, and reconnaissance
ITU	International Telecommunication Union
KPN	Koninklijke PTT Nederland
LMT	Latvijas Mobilais Telefons
MitM	man in the middle
MTS	Mobile TeleSystems
NATO	North Atlantic Treaty Organization
OUSD R&E	Office of the Under Secretary of Defense for Research and Engineering

REB	radioelectronic struggle (radioelektronnaia bor'ba)
RSOI	reception, staging, onward movement, and integration
SME	subject-matter expert
SPOD	seaport of debarkation
TAA	tactical assembly area
TS	Technical Specification
TTP	tactics, techniques, and procedures
UAV	unmanned aerial vehicle
USAREUR	U.S. Army Europe

References

"3.4–3.8 GHz Auctions in Finland," European 5G Observatory, October 2, 2018.

3GPP—*See* 3rd Generation Partnership Project.

3rd Generation Partnership Project, *Service Requirements for the 5G System (3GPP TS 22.261 version 17.10.0 Release 17)*, Sophia Antipolis, France: ETSI, May 2022.

5G-Blueprint, homepage, undated. As of May 24, 2022:
https://www.5gblueprint.eu/

5G Infrastructure Public Private Partnership, "5G-MOBIX," webpage, undated-a. As of May 24, 2022:
https://5g-ppp.eu/5g-mobix/

———, "About the 5G PPP," webpage, undated-b. As of March 28, 2022:
https://5g-ppp.eu/

———, "Development of the 5G Infrastructure PPP in Horizon 2020," webpage, undated-c. As of March 28, 2022:
https://5g-ppp.eu/history/

5G PPP—*See* 5G Infrastructure Public Private Partnership.

5GRAIL, homepage, undated. As of May 24, 2022:
https://5grail.eu/

5G-ROUTES, homepage, undated. As of May 24, 2022:
https://www.5g-routes.eu/

Abdallah, Ali A., Kimia Shamaei, and Zaher M. Kassas, "Assessing Real 5G Signals for Opportunistic Navigation," *Proceedings of the 33rd International Technical Meeting of the Satellite Division of the Institute of Navigation (ION GNSS+ 2020)*, September 2020, pp. 2548–2559.

Alotaibi, Daifallah, "Survey on Network Slice Isolation in 5G Networks: Fundamental Challenges," *Procedia Computer Science*, Vol. 182, 2021, pp. 38–45.

Army Techniques Publication 3-35, *Army Deployment and Redeployment*, Washington, D.C.: Headquarters, U.S. Department of the Army, March 2015.

Authority for Consumers & Markets, *5G and the Netherlands Authority for Consumers and Markets*, The Hague, The Netherlands, December 12, 2018.

Axon Partners Group Consulting, *Study on the Financial and Environmental Impact of 5G and a Fourth Mobile Network Operator in Belgium*, Brussels: Belgian Institute for Postal Services and Telecommunications, April 2021.

Barton, James, "Bite Pledges 5G Investment as It Switches on New Sites," Developing Telecoms, January 21, 2021a.

———, "Estonia to Put Three 3.5GHz Licences Up for Auction Next Year," Developing Telecoms, December 17, 2021b.

———, "All Available Spectrum Sold in Latvia's 700MHz Auctions," Developing Telecoms, January 4, 2022.

"Belgian 5G Auction Terms Receive Government Approval," CommsUpdate, November 25, 2021.

Belgian Institute for Postal Services and Telecommunications, "5G," webpage, undated-a. As of March 28, 2022:
https://www.bipt.be/consumers/5g

———, "BIPT," webpage, undated-b. As of May 24, 2022:
https://www.bipt.be/consumers/bipt

———, *Communication by the BIPT Council of 10 September 2018 Regarding the Introduction of 5G in Belgium*, Brussels, September 10, 2018.

———, *Annual Report 2019*, Brussels, May 29, 2020.

Busvine, Douglas, "Germany Raises 6.55 Billion Euros in Epic 5G Spectrum Auction," Reuters, June 12, 2019.

"Camouflage" ["Maskirovka"], *Strategic Rocket Forces Encyclopedia* [*Entsiklopediia RVSN*], Ministry of Defense of the Russian Federation, undated.

Daly, John C. K., "Lithuanian-Russian Radio Frequency Dispute Highlights Problems of Civilian Versus Military Applications," *Eurasia Daily Monitor*, Vol. 18, No. 38, March 8, 2021.

"DNA's 5G Network Coverage Now Reaches More than 3.5 Million Finns in Nearly 140 Towns as a Consequence of a New 5G Frequency Band," DNA, December 13, 2021.

Dobrynin, Sergei, and Mark Krutov, "Communication Breakdown: How Russia's Invasion of Ukraine Bogged Down," RFERL, March 19, 2022.

DoD—*See* U.S. Department of Defense.

Dura, Maksymilian, "Electronic Warfare: Russian Response to the NATO's Advantage? [ANALYSIS]," Defence24, May 5, 2017.

"Dutch Multi-Band 5G Auction Raises EUR1.23bn," CommsUpdate, July 22, 2020.

Dziadul, Chris, "First 5G Auction for Lithuania," Broadband TV News, October 26, 2021.

Elisa, "CEO's Review Q4 2021," webpage, undated. As of March 28, 2022:
https://elisa.com/corporate/investors/elisa-as-an-investment/ceos-review/

———, "Elisa First in World to Launch Commercial 5G," press release, Helsinki, June 27, 2018.

Ericsson, "Lithuania Added to Telia and Ericsson Nordics and Baltics 5G Alliance," press release, Stockholm, November 3, 2020.

European 5G Observatory, "What Is the European 5G Observatory?" webpage, undated. As of March 28, 2022:
https://5gobservatory.eu/about/what-is-the-european-5g-observatory/

———, "Estonia," webpage, last updated 2021a. As of March 28, 2022:
https://5gobservatory.eu/national-5g-spectrum-assignment/#1533308955240-d7784498-af7e

———, "National 5G Strategies: Belgium," webpage, last updated 2021b. As of June 14, 2022:
https://5gobservatory.eu/national-5g-plans-and-strategies/#1533564498506-0ca6acde-c41f

———, "National 5G Strategies: Poland," webpage, last updated 2021c. As of March 28, 2022:
https://5gobservatory.eu/national-5g-plans-and-strategies/#1533566084406-4e0a1da3-0827

———, "To Fulfil Its Potential 5G Needs Access to Much Higher Frequencies: 3.5 GHz and Above. This Was Not the Case with Earlier Mobile Generations," webpage, last updated 2021d. As of June 14, 2022:
https://5gobservatory.eu/5g-spectrum/#:~:text=
Internationally%20there%20has%20been%20a,(24.25%2D27.5%20GHz)

European Commission, *The 5G Infrastructure Public Private Partnership (5G PPP): First Wave of Research & Innovation Projects*, Brussels, undated-a.

———, *Digital Economy and Society Index (DESI) 2021: Belgium*, Brussels, undated-b.

———, *Digital Economy and Society Index (DESI) 2021: DESI Methodological Note*, Brussels, undated-c.

———, *Digital Economy and Society Index (DESI) 2021: Estonia*, Brussels, undated-d.

———, *Digital Economy and Society Index (DESI) 2021: Finland*, Brussels, undated-e.

———, *Digital Economy and Society Index (DESI) 2021: Germany*, Brussels, undated-f.

———, *Digital Economy and Society Index (DESI) 2021: Latvia*, Brussels, undated-g.

———, *Digital Economy and Society Index (DESI) 2021: Lithuania*, Brussels, undated-h.

———, *Digital Economy and Society Index (DESI) 2021: Netherlands*, Brussels, undated-i.

———, *Digital Economy and Society Index (DESI) 2021: Norway*, Brussels, undated-j.

———, *Digital Economy and Society Index (DESI) 2021: Poland*, Brussels, undated-k.

———, *Digital Economy and Society Index (DESI) 2021: Sweden*, Brussels, undated-l.

———, "What We Do—Communications Networks, Content and Technology: Mission Statement of the Directorate-General for Communications Networks, Content and Technology (Connect)," webpage, undated-m. As of March 28, 2022: https://ec.europa.eu/info/departments/communications-networks-content-and-technology/what-we-do-communications-networks-content-and-technology_en

———, *Communication from the Commission to the European Parliament, the Council, the European Economic and Social Committee and the Committee of the Regions—5G for Europe: An Action Plan*, Brussels, September 14, 2016.

———, "European Commission to Harmonise the Last Pioneer Frequency Band Needed for 5G Deployment," press release, Brussels, May 14, 2019.

———, *Communication from the Commission to the European Parliament, the Council, the European Economic and Social Committee and the Committee of the Regions, 2030 Digital Compass: The European Way for the Digital Decade*, Brussels, March 9, 2021.

———, "Broadband in Estonia," webpage, last updated February 4, 2022a. As of May 24, 2022: https://digital-strategy.ec.europa.eu/en/policies/broadband-estonia

———, "5G Cross-Border Corridors," webpage, last updated February 22, 2022b. As of March 24, 2022: https://digital-strategy.ec.europa.eu/en/policies/cross-border-corridors

———, "5G Action Plan," webpage, last updated February 24, 2022c. As of March 28, 2022: https://digital-strategy.ec.europa.eu/en/policies/5g-action-plan

———, "5G Observatory," webpage, last updated February 24, 2022d. As of March 28, 2022: https://digital-strategy.ec.europa.eu/en/policies/5g-observatory

———, "Connected and Automated Mobility," webpage, last updated February 24, 2022e. As of March 28, 2022: https://digital-strategy.ec.europa.eu/en/policies/connected-and-automated-mobility

European Commission Directorate-General for Communications Networks, Content and Technology, *5G Observatory Quarterly Report 13 Up to October 2021*, Brussels, October 2021.

European External Action Service, *EU Concept for Reception, Staging, Onward Movement and Integration (RSOI) for EU-Led Military Operations*, Brussels: Council of the European Union, EEAS 0078/12, May 11, 2012.

European Parliamentary Research Service, "Example: Estonia," webpage, undated. As of March 28, 2022:
https://map.sciencemediahub.eu/5g#m=4/1401.69686/48.86747,p=37

"The European Part of Russia May Be Left Without 5G Communications" ["Evropaiskaia chast' Rossii mozhet ostat'sia bez sviazi 5G"], *Russia Today PRIME* [*Rossiia Segodnia PRAIM*], October 14, 2020.

Federal Ministry of Transport and Digital Infrastructure (Germany), *5G Strategy for Germany: A Scheme to Promote the Development of Germany to Become a Lead Market for 5G Networks and Applications*, Berlin, 2017.

Field Manual 4-01, *Army Transportation Operations*, Washington, D.C.: U.S. Department of the Army, April 2014.

George, David, Dennisa Nichiforov-Chuang, and Emanuel Kolta, *How Spectrum Will Shape the Outlook for 5G in Russia*, London: GSMA, 2020.

"German 5G Spectrum Auction Concludes," HBR Radiofrequency Technologies, June 12, 2019.

"German Spectrum Auction Raises EUR5.08bn," CommsUpdate, June 22, 2015.

Giles, Keir, "Assessing Russia's Reorganized and Rearmed Military," white paper, Carnegie Endowment for International Peace, May 3, 2017.

Gilles, François, and Jaroslav Toth, *Accelerating the 5G Transition in Europe: How to Boost Investments in Transformative 5G Solutions*, Brussels: European Commission, February 2021.

Goudos, Sotirios K., Panagiotis I. Dallas, Stella Chatziefthymiou, and Sofoklis Kyriazakos, "A Survey of IoT Key Enabling and Future Technologies: 5G, Mobile IoT, Sematic Web and Applications," *Wireless Personal Communications*, Vol. 97, 2017, pp. 1645–1675.

"Government Approves the Use of 24 GHz Radio Frequency Band for 5G Networks" ["Pravitel'stvo utverdilo ispol'zovanie radiochastot 24 GGts dlia seti 5G"], Interfax, May 7, 2021.

Government Offices of Sweden, *A Completely Connected Sweden by 2025—a Broadband Strategy*, Stockholm, undated.

Grassi, Paul A., Michael E. Garcia, and James L. Fenton, *NIST Special Publication 800-63-3: Digital Identity Guidelines*, Gaithersburg, Md.: National Institute of Standards and Technology, June 2017.

Grove, Thomas, Julian E. Barnes, and Drew Hinshaw, "Russia Targets NATO Soldier Smartphones, Western Officials Say," *Wall Street Journal*, October 4, 2017.

Guzdar, Maya, and Tomas Jermalavičius, "Between the Chinese Dragon and American Eagle: 5G Development in the Baltic States," brief, Tallinn: International Centre for Defence and Security, August 2020.

Guzenko, V. F., and A. L. Moraresku, "Radioelectronic Struggle: Contemporary Substance" ["Radioelektronnaia bor'ba: Sovremennoe soderzhanie"], *Radioelectronic Struggle in the Armed Forces of the Russian Federation* [*Radioelektronnaia bor'ba v Vooruzhennykh Silakh Rossiiskoi Federatsii*], 2017.

HBR Radiofrequency Technologies, "Finland," webpage, undated-a. As of March 28, 2022: https://halberdbastion.com/intelligence/countries-nations/finland

———, "Russia," webpage, undated-b. As of June 29, 2021: https://halberdbastion.com/intelligence/countries-nations/russia

———, "Tele2 Russia," webpage, undated-c. As of April 25, 2021: https://halberdbastion.com/intelligence/mobile-networks/tele2-russia

———, "Telekom Deutschland," webpage, undated-d. As of March 28, 2022: https://halberdbastion.com/intelligence/mobile-networks/telekom-deutschland

———, "Vodafone Germany," webpage, undated-e. As of March 28, 2022: https://halberdbastion.com/intelligence/mobile-networks/vodafone-germany

Iastrebova, Svetlana, "Putin Doesn't Hand Over Popular 5G Frequencies to Mobile Service Providers" ["Putin ne otdaet operatoram popularnye chastoty za 5G"], *Vedomosti*, August 15, 2019.

"Independent 'Tower' Operators Are Headed for the Regions" ["Nezavisimye 'bashennye' operatory poidut v regiony"], IKS Media, December 6, 2018.

"Inmarsat Wins Injunction Against Dutch Refarming of 3.5 GHz Band," Telecompaper, June 30, 2021.

International Telecommunication Union Radiocommunication Sector, *Minimum Requirements Related to Technical Performance for IMT-2020 Radio Interface(s)*, Geneva, Switzerland, M.2410-0, November 2017.

Joint Publication 3-35, *Deployment and Redeployment Operations*, Washington, D.C.: U.S. Joint Chiefs of Staff, January 2018.

Joint Publication 3-85, *Joint Electromagnetic Spectrum Operations*, Washington, D.C.: U.S. Joint Chiefs of Staff, May 22, 2020.

Karwala, Damian, "5G Regulation and Law in Poland," CMS, undated.

Khan, Latif U., Ibrar Yaqoob, Nguyen H. Tran, Zhu Han, and Choong Seon Hong, "Network Slicing: Recent Advances, Taxonomy, Requirements, and Open Research Challenges," *IEEE Access*, Vol. 8, 2020, pp. 36009–36028.

Kharpukhin, V. I., "On the 100th Anniversary of the Birth of the Formidable Scholar in the Theory and Practice of Electronic Warfare V. I. Kuznetsov" ["K 100-letiiu so dnia rozhdeniia krupnogo uchenogo v oblasti teorii i praktiki radioelectronnoi bor'by V. I. Kuznetsova"], *Military Thought* [*Voennaia Mysl'*], No. 12, 2020, pp. 108–115.

Kiniakina, Ekaterina, "Russian Mobile Service Providers May Miss Out on the Possibility of Building 5G Networks" ["Rossiiskie operatory mozhet lishit'sia vozmozhnostistroit' seti 5G"], *Vedomosti*, September 20, 2020.

Kjellén, Jonas, *Russian Electronic Warfare: The Role of Electronic Warfare in the Russian Armed Forces*, Stockholm: Swedish Defence Research Agency, FOI-R—14625—SE, September 2018.

Kodachigov, Valery, "'Megafon' and 'Rostelekom' Get Frequencies for Fifth-Generation Communications" ["'Megafon' i 'Rostelekom' nashli chastoty dlia piatogo pokoleniia sviazi"], *Vedomosti*, September 20, 2018.

———, "Russian 5G Interferes with NATO Airplanes" ["Rossiiskii 5G meshaet samoletom NATO"], *Vedomosti*, February 7, 2021.

Kondakov, M. S., and V. V. Solopov, "The Potential of LTE/5G Mobile Technologies in Military Communications" ["Vozmozhnosti mobil'nykh tekhnologii LTE/5G v voennoi sviazi"], *Theory and Practice of Radio Communications* [*Teoriia i tekhnika radiosviazi*], No. 4, 2020, pp. 36–47.

Kopera, Dominik, "'5G for Poland' Strategy," briefing slides, Warsaw: Ministry of Digital Affairs, undated.

Korolyov, I., S. Kozlitin, and O. Nikitin, "Problems of Determining Ways of Employing Forces and Means of Electronic Warfare" ["Problemy opredeleniya sposobov boevogo primeneniya sil i sredstv radioelektronnoy bor'by"], *Military Thought* [*Voennaia Mysl'*], No. 9, 2016, pp. 14–19.

"KPN, T-Mobile and VodafoneZiggo Acquire Frequencies in Dutch Mobile Communications Auction," Government of the Netherlands, July 21, 2020.

Krieger, Jörn, "Germany Opens 700 MHz Frequency Auction," Broadband TV News, January 29, 2015.

Lastochkin, I. A., "Role and Place of Electronic Warfare in Contemporary and Future Combat Actions" ["Rol' i mesto radioelektronnoy bor'by v sovremennykh i budushchikh boyevykh deystviyakh"], *Military Thought* [*Voennaia Mysl'*], No. 12, 2015, pp. 14–19.

81

Lastochkin, I. A., I. E. Donskov, and A. L. Moraresku, "An Analysis of Contemporary Conceptions for Undertaking Operations in the Electromagnetic Spectrum from the Standpoint of Radioelectronic Struggle" ["Analiz covremmennykh kontseptsii po vedeniiu operatsii v elektromagnitnom spektre s pozitsii radioelektronnoi bor'by"], *Military Thought* [*Voennaia Mysl'*], No. 4, 2021, pp. 29–38.

"Latvian Mobile Communications Operators Pay a Pretty Penny for 5G Frequency Rights," Baltic News Network, December 27, 2021.

Lee, Jaehun, Ji-Seon Paek, and Songcheol Hong, "Frequency Reconfigurable Dual-Band CMOS Power Amplifier for Millimeter-Wave 5G Communications," *2021 IEEE MTT-S International Microwave Symposium (IMS)*, 2021, pp. 846–849.

Lennighan, Mary, "Norway to Get More Competitive as Ice Steps Up," Telecoms, November 18, 2021.

Lidy, A. Martin, Douglas P. Baird, John M. Cook, Robert C. Holcomb, Samuel H. Packer, and William J. Sheleski, *Doctrine, Organizations, and Systems for Reception, Staging, Onward Movement, and Integration (RSOI) Operations*, Alexandria, Va.: Institute for Defense Analyses, January 1997.

"Lithuania Approved the 5G Development Guidelines for 2020–2025," Lithuanian Ministry of Transport and Communications, June 3, 2020.

"Lithuania Bans 'Unreliable' Technologies from Its 5G Network," LRT English, May 25, 2021.

"Lithuania's 5G Development Hampered by Russian Military Infrastructure," LRT English, February 23, 2021.

"LMT Deploys 100th 5G Base Station," CommsUpdate, December 10, 2021.

"LMT Flicks the Switch—Becomes First Mobile Operator to Launch 5G Internet in the Baltics," *Baltic Times*, July 26, 2019.

"LMT Secures 5G-Compatible Spectrum," CommsUpdate, December 11, 2017.

"LMT Will Install 100 5G Base Stations This Year," LMT, September 8, 2021.

"LMT Will Install 100 5G Base Stations This Year," Labs of Latvia, October 22, 2021.

Makarenko, S. I., *Information Confrontation and Radioelectronic Struggle in the Network-Centric Wars of the Early 21st Century* [*Informatsionnoe protivoborstvo i radioelektronnaia bor'ba v setetsentricheskikh voinakh nachala XXI veka*], Saint Petersburg, Russia: Naukoemkie Tekhnologii, 2017.

Mann, Colin, "Finland Concludes 26 GHz 5G Spectrum Auction," Advanced Television, June 9, 2020.

McCabe, Thomas R. "The Russian Perception of the NATO Aerospace Threat: Could It Lead to Preemption?" *Air & Space Power Journal*, Fall 2016, pp. 64–77.

McDermott, Roger N., *Russia's Electronic Warfare Capabilities to 2025: Challenging NATO in the Electromagnetic Spectrum*, Tallinn: International Centre for Defence and Security, 2017.

McMorrow, Ryan, Anna Gross, Polina Ivanova, and Kathrin Hille, "Huawei Faces Dilemma over Russia Links That Risk Further US Sanctions," *Ars Technica*, April 1, 2022.

Ministry of Defence of the Russian Federation, *Conceptual Views on the Responsibilities of the Armed Forces of the Russian Federation in the Information Space* [*Kontseptual'nye vzgliady na deiatel'nost' Vooruzhennykh Sil Rossiiskoi Federatsii v informatsionnom prostranstve*], Moscow, 2011.

Ministry of Economic Affairs and Climate Policy, *Connectivity Action Plan*, The Hague, The Netherlands, July 2018.

Ministry of Transport and Communications, *Turning Finland into the World Leader in Communications Networks—Digital Infrastructure Strategy 2025*, Helsinki, 2019.

Mohyeldin, Eiman, "Minimum Technical Performance Requirements for IMT-2020 Radio Interface(s)," presented at the ITU-R Workshop on IMT-2020 Terrestrial Radio Interfaces Evaluation, Geneva, Switzerland, December 2019.

Monserrat, Jose F., Genevieve Mange, Volker Braun, Hugo Tullberg, Gerd Zimmermann, and Ömer Bulakci, "METIS Research Advances Towards the 5G Mobile and Wireless System Definition," *EURASIP Journal on Wireless Communications and Networking*, No. 53, 2015.

Morris, Anne, "VodafoneZiggo Uses Dynamic Spectrum Sharing to Get Ahead with 5G," Light Reading, April 28, 2020.

———, "Tele2 Picks Nokia for 5G RAN in Baltics," Light Reading, January 4, 2022a.

———, "Telia Gets 5G Ball Rolling in Lithuania with DSS," Light Reading, January 14, 2022b.

"MTS Rolls Out Pilot Consumer 5G Network in Moscow" ["MTS zapustit v Moskve pol'zovatel'skuiu pilotnuiu set' 5G"], Interfax, March 5, 2021.

Naglis, Alla Y., and Xenia A. Melkova, "Telecoms in Russia," King & Spalding LLP, April 29, 2019.

Nin, Catherine Sbeglia, "Telia Turns on 5G in Estonia with Ericsson," RCR Wireless, November 11, 2020.

"Nkom Announces Results of 700MHz, 2100MHz Spectrum Auction," CommsUpdate, June 6, 2019.

"Nkom Announces Results of 700MHz, 2100MHz Spectrum Auction in Norway," European 5G Observatory, June 14, 2019.

Nokia, "Nokia Wins Five-Year 5G Deal with Elisa Estonia as Sole RAN Vendor," press release, Espoo, Finland, December 20, 2021.

Norwegian Communications Authority, "The Norwegian 5G Auction Has Concluded," webpage, September 30, 2021. As of May 24, 2022: https://www.nkom.no/aktuelt/the-norwegian-5g-auction-has-concluded

Oanh, Van, "Gmobile Replaces Beeline, Targets Mekong," *Saigon Times*, September 19, 2012.

O'Dea, S., "Estimated Share of Population Covered by at Least LTE or WiMAX Mobile (Cellular) Network Worldwide and in Rural and Urban Areas in 2021, by Region," Statista, December 8, 2021.

O'Dwyer, Gerard, "Finland, Norway Press Russia on Suspected GPS Jamming During NATO Drill," Defense News, November 16, 2018.

"Orange Launches 5G Network," ICT Market Experts, June 5, 2020.

Osborne, Charlie, "4G, 5G Networks Could Be Vulnerable to Exploit Due to 'Mishmash' of Old Technologies," ZDNet, October 1, 2020.

"Plus Extends 5G Coverage. Over 7 Million of Poles Within the 5G Network Coverage," Grupa Polsat Plus, January 5, 2021.

"Polish Senate Approves Delay to 700MHz Switchover," CommsUpdate, March 1, 2019.

Polityuk, Pavel, and Jim Finkle, "Ukraine Says Communications Hit, MPs Phones Blocked," Reuters, March 4, 2014.

President of Russia, *Military Doctrine of the Russian Federation* [*Voennaia doktrina Rossiiskoi Federatsii*], Moscow, December 26, 2014.

Promixus, "Frequently Asked Questions About 5G," webpage, undated-a. As of March 28, 2022: https://www.proximus.be/support/en/id_sfaql_5g_global_me_cor/large-companies/support/telephony/mobile-phone-and-sim-card/set-up-your-mobile-phone/frequently-asked-questions-about-5g.html

———, "Launching 5G in Belgium," fact sheet, undated-b. As of March 28, 2022: https://www.proximus.be/dam/jcr:55533eab-0735-47c9-a971-d40003eec3c7/cdn/sites/iportal/documents/pdfs/cor/5g-development-of-4g/200625_5G-infographic_EN_CTA~2020-12-10-15-15-10~cache.pdf

Pujol, Frédéric, Carole Manero, Basile Carle, and Santiago Remis, *5G Observatory Quarterly Report 12: Up to June 2021*, Brussels: European Commission, July 2021.

Qiang Chen, Xiaolei Wang, and Yingying Lv, "An Overview of 5G Network Slicing Architecture," *AIP Conference Proceedings*, Vol. 1967, No. 1, 2018.

"R-330Zh Zhitel Russian Cellular Jamming and Direction Finding System," webpage, undated. As of March 28, 2022:
https://odin.tradoc.army.mil/mediawiki/index.php/R-330Zh_Zhitel_Russian_Cellular_Jamming_and_Direction_Finding_System

Republic of Estonia Ministry of Economic Affairs and Communications, *Digital Agenda 2020 for Estonia: Updated 2018 (Summary)*, Tallinn, 2019.

Robles-Carrillo, Margarita, "European Union Policy on 5G: Context, Scope and Limits," *Telecommunications Policy*, Vol. 45, No. 8, 2021.

Samczyński, Piotr, Karol Abratkiewicz, Marek Płotka, Tomasz P. Zieliński, Jacek Wszołek, Sławomir Hausman, Piotr Korbel, and Adam Księżyk, "5G Network-Based Passive Radar," *IEEE Transactions on Geoscience and Remote Sensing*, Vol. 60, No. 5108209, 2022.

Secretary of State for Economic Affairs and Climate Policy, Minister of Justice and Security, and State Secretary of the Interior and Kingdom Relations, *Dutch Digitalisation Strategy 2018–2021: Getting the Netherlands Ready for the Digital Future*, The Hague, The Netherlands, September 30, 2019.

———, *The Dutch Digitalisation Strategy 2021*, The Hague, The Netherlands, June 22, 2021.

Shaik, Altaf, Ravishankar Borgaonkar, Shinjo Park, and Jean-Pierre Seifert, "New Vulnerabilities in 4G and 5G Cellular Access Network Protocols: Exposing Device Capabilities," *Proceedings of the 12th Conference on Security and Privacy in Wireless and Mobile Networks*, May 2019, pp. 221–231.

Simerly, Mark T., *Improving Reception, Staging, Onward Movement, and Integration Operations for the Interim and Objective Forces*, thesis, Fort Leavenworth, Kan.: U.S. Army Command and General Staff College, 2002.

Sitronics, "Telecommunications" ["Telekomunikatsii"], webpage, undated. As of June 29, 2021:
https://www.sitronics.com/businesses?industries=telecom

Smith, Alexander, "Norway Calling Out Russia's Jamming Shows European Policy Shift," NBC News, November 24, 2018.

Song, Julie, "Why Low Latency (Not Speed) Makes 5G a World-Changing Technology," *Forbes*, February 6, 2020.

"Source: Minkomsviaz and Mobile Service Providers Oriented Toward Frequencies from 3.4–3.8 GHz for 5G Development" ["Istochnik: Minkomsviaz' i operatory orientiruiutsia na chastoty 3.4–3.8 GGts dlia razvitiia 5G"], *TASS*, August 15, 2019.

Swedish Post and Telecom Authority, "Assignment in the 700 MHz Band," webpage, December 11, 2018. As of March 28, 2022:
https://www.pts.se/en/english-b/radio/auctions/700/

———, "Assignment in the 3.5 GHz and 2.3 GHz Bands," webpage, last updated January 21, 2021. As of March 28, 2022:
https://www.pts.se/en/english-b/radio/auctions/assignment-in-the-3.4---3.8-ghz-bandet/

Tele2, "Tele2 5G Is Here," webpage, undated. As of March 28, 2022:
https://www.tele2.com/about/what-we-offer/tele2-5g/#:~:text=Tele2%20launched%20its%20first%20public,a%20more%20energy%20efficient%20way.

———, "Tele2 Wins Latvian 5G Frequencies," press release, Stockholm, December 16, 2021.

———, "Tele2 Chooses Nokia for 5G RAN Rollout in the Baltics," press release, Stockholm, January 4, 2022.

"Tele2 and Bite Claim Success in 700MHz Auction Ahead of Final Approval of Results," CommsUpdate, December 20, 2021.

"Tele2 Latvia Snaps Up 3.5GHz Spectrum for 5G," CommsUpdate, September 17, 2018.

Telefónica, "Focus on 'Pure 5G' over 3.6 GHz Frequency Fast O_2 Network Expansion: 50 Per Cent 5G Coverage by the End of 2022," press release, Munich, Germany, October 5, 2021.

"Telenet to Launch 5G in Belgium Using Provisional Spectrum," European 5G Observatory, December 8, 2021.

"Telenor to Provide 5G in All 11 Norwegian Counties in 2022, Targeting 250 New Sites," Telecompaper, September 30, 2021.

Telia Company, "Telia Sweden's First Major 5G Network Up and Running in Stockholm," press release, Solna, Sweden, May 24, 2020.

———, "The Need for Speed: Telia and Ericsson Beat Swedish 5G Speed Record," press release, Solna, Sweden, July 5, 2021a.

———, "Telia Company Launches 5G SA Core Network in Finland—the First Commercially Available 5G SA Network in the Nordic and Baltic Region," press release, Solna, Sweden, November 10, 2021b.

"Telia Is Launching 5G Network in Lithuania," Telia Company, November 10, 2020.

"Telia Is Ready for Commercial 5G Launch: Installed 250 Base Stations," Telia Company, October 13, 2021.

"Telia Launches 5G in Norway," Telia Company, May 12, 2020.

"Telia Launches First Public 5G Network in Estonia," Telia Company, November 10, 2020.

"Telia Lietuva Launches Trial 5G Network," CommsUpdate, November 11, 2020.

"Telia Norge's 5G Network Now Covers 30% of the Population," CommsUpdate, September 22, 2021.

"Telia's 5G Network Available at over 50 Locations in Estonia," *Baltic Times*, June 14, 2021.

Thomas, Timothy, *Russia's Electronic Warfare Force: Blending Concepts with Capabilities*, McLean, Va.: MITRE Center for Technology and National Security, September 10, 2020.

"T-Mobile 5G Covers 25% of Poland," CommsUpdate, September 16, 2021.

"T-Mobile First Dutch Telecom to Offer Nationwide 5G Coverage," NL Times, October 27, 2020.

Tomás, Juan Pedro, "Three Sweden Launches 5G in Six Cities," RCR Wireless, June 16, 2020a.

———, "T-Mobile Netherlands, KPN to Launch Commercial 5G Networks Next Week," RCR Wireless, July 24, 2020b.

———, "Telenor Launches 5G in Stockholm, Targets Nationwide Coverage by 2023," RCR Wireless, October 30, 2020c.

———, "Telefonica Deutschland Reaches 80 German Cities with 5G Technology," RCR Wireless, July 6, 2021a.

———, "Vodafone Germany Expands 5G Footprint, Adds New 5G Devices," RCR Wireless, August 13, 2021b.

———, "Telenet to Start Gradual Deployment of 5G in Belgium Next Week," RCR Wireless, December 3, 2021c.

———, "Vodafone Reaches over 45 Million People in Germany with Its 5G Network," RCR Wireless, January 13, 2022a.

———, "Deutsche Telekom's 5G Network Covers over 90% of Germany's Population," RCR Wireless, February 16, 2022b.

Traficom Finnish Transport and Communications Agency, "Spectrum Auction for 26 GHz Frequency Band," webpage, June 11, 2020. As of March 28, 2022: https://www.traficom.fi/en/communications/communications-networks/spectrum-auction-26-ghz-frequency-band

Trevithick, Joseph, "Russia Jammed Phones and GPS in Northern Europe During Massive Military Drills," The Warzone, October 16, 2017.

Turk, Hala, "Telia Eesti 5G Network Reaches 77 Regions in Estonia," Inside Telecom, July 26, 2021.

U.S. Department of Defense, *Electromagnetic Spectrum Strategy*, Washington, D.C., September 2013.

———, *Electromagnetic Spectrum Superiority Strategy*, Washington, D.C., October 2020.

van der Klugt, Geert, "Orange Belgium Activates 5G in Antwerp, Rest of Belgium to Follow," Techzine, February 11, 2022.

"Vodafone First to Offer Nationwide 5G," Vodafone, July 20, 2020.

ZALA Aero Group, "ZALA Aero," webpage, undated. As of March 28, 2022:
https://zala-aero.com/en/about/

"Zala Aero Group REX-1 Drone Jammer," Defense Update, August 26, 2018.

Zetter, Kim, "The Critical Hole at the Heart of Our Cell Phone Networks," *Wired*, April 28, 2016.

Zhriblis, Andrei, "Is a Federal Antimonopoly Service of Russia Decision Nearing About 5G?" ["Priblizit li reshenie FAS Rossiiu k 5G"], BFM.RU, May 4, 2021.